Organic Management for the Professional

Organic Management for the Professional

The Natural Way for Landscape Architects and Contractors, Commercial Growers, Golf Course Managers, Park Administrators, Turf Managers, and Other Stewards of the Land

Howard Garrett, John Ferguson, and Mike Amaranthus

University of Texas Press Austin

Printed in the United States of America
First edition, 2012

Requests for permission to reproduce material from this work should be sent to:
Permissions
University of Texas Press
P.O. Box 7819
Austin, TX 78713–7819
www.utexas.edu/utpress/about/bpermission.html

The paper used in this book meets the minimum requirements of ANSI/NISO Z39.48–1992 (R1997) (Permanence of Paper).

Library of Congress Cataloging-in-Publication Data

Garrett, Howard, 1947–
Organic management for the professional : the natural way for landscape architects and contractors, commercial growers, golf course managers, park administrators, turf managers, and other stewards of the land / Howard Garrett, John Ferguson, Mike Amaranthus. — 1st ed.
p. cm.
Includes bibliographical references and index.
ISBN 978-0-292-72921-6 (cloth : alk. paper) — ISBN 978-0-292-73712-9 (e-book)
1. Organic gardening. 2. Landscape gardening. I. Ferguson, John, 1940– II. Amaranthus, Michael P. III. Title.
SB453.5.G39 2012
635.9'87—dc23 2011042851

Contents

Preface

We're All in This Together

This book is for landscape architects and contractors, commercial growers, golf course managers, park administrators, turf managers, and other stewards of the land. It is also for those in the position of hiring these people.

The organic approach is about a different way of thinking about landscaping, farming, gardening, pest control, and living in general. It's not just about switching products. It's about reading and studying and listening to get a better understanding of nature so you can make better decisions, grow better plants, improve our health, reduce costs, and be profitable, or at least stay within the budget if you are a nonprofit entity. Organic programs on a large scale can make a difference in the quality of our environment, our use of fossil fuels, and our climate. Let's face it: We are in an era of declining, more expensive resources. Whether it is water, fertilizer, or fossil fuels, the organic approach taken to the large scale has enormous potential to conserve resources and create a more sustainable future for everyone. After all, we are all in this together.

Critics say, "Organic programs might work in residential gardens, but they won't work on a large scale. They cost too much, don't control the pests, and are too much work." Food producers and many academics say, "We have to feed the world." Landscape managers say, "It's impractical." The truth is that natural organic programs work better in every way. They are easier to do on a large or commercial scale than on a small scale. Successful organic programs exist on all scales today, all over the world, although they aren't the mainstream yet. The United States has certified organic farms encompassing over one million acres and organic food sales totaling more than $7 billion per year. Despite this, universities and agricultural extension agencies keep preaching that organics won't work and that the synthetic chemical approach is the only answer. People can't be kept in the dark much longer.

Being organic on a commercial scale is no different from being organic in the backyard garden—just bigger. Yes, economics is a major factor. You'll learn that the front end of a project, during the soil-building stage, is the only possible time an organic program is more expensive. Those costs are not wasted, however, as they are with the synthetic approach. Expenses for materials and actions that build and improve the health of the soil are investments that pay off in the long term. These expenses also have immediate benefits for our environment by reducing erosion, conserving water, and sequestering carbon. In contrast, synthetic fertilizers and pesticides hurt the natural systems, and the damage accumulates with every application of toxic materials. The natural organic method is without doubt the more cost-efficient way to go. Properly designed organic programs can be switched to cold turkey at no more cost than the toxic chemical approach.

Natural organic products build the soil organic matter, stimulate life in the soil, and cause the soil to get better every year—forever. Toxic pesticides and high-nitrogen synthetic fertilizers, on the other hand, hurt the soil and pollute the environment. Every application reduces soil health until, one day, total failure occurs and replanting is needed. The end result is the destruction of the soil. The common explanation is "worn-out soil," which means a "dead soil" condition that is a self-inflicted casualty brought on by improper management. Conventional synthetic chemical farming and other growing operations are not sustainable. They waste money and continually deplete the soil of humus, water-holding capacity, mineral nutrients, beneficial life, and production capacity.

The natural organic method on a large scale is the only answer. Nature's method was in place and functioning perfectly on a global scale long before humans appeared on Earth. We can continue to make better and better decisions to improve the health of the soil, plants, pets, livestock, and people, but the basic program is just common sense. Natural organic horticulture and agriculture programs are productive and sustainable and need to be a big part of our future.

From the food we eat, to the air we breath, to the clothes we wear, humans depend on the thin covering of the earth's surface we call soil. Arguably, this thin and fragile layer of living topsoil is the earth's most critical natural resource. The quality of the soil determines, to a great degree, the condition of our air, water, and food, as well as the overall quality of all life on Earth. As you see later in this book, it is also a critical factor in the fight against global warming. Yet to most, soil is just plain dirt, understood by few and taken for granted by most. This book, however, is a celebration of soil and is focused on reversing soil destruction and building healthy soil.

After only a half century of conventional agriculture using high-analysis synthetic fertilizers, the organic content of most farmlands has fallen way below a sustainable level, in some cases as low as one-tenth its original content. With the loss of the organic matter, rainwater or irrigation can no longer penetrate the soil fast enough to prevent it from running off, causing floods carrying topsoil and nutrients to the ocean. Loss of productive soil has created the so-called need for pesticides. Each year, more and more of these poisons are used. We are now using over 2 billion pounds per year in the United States. Pests are damaging more of our plants now than before modern artificial pesticides were invented. Toxic pesticides are good for one thing: keeping unhealthy plants alive. And it all keeps going downhill. Unhealthy plants affect the health of animals and people. Just as we have all participated in creating these problems, we all need to work together to find and implement solutions. The solution to many of the health issues in the world is organic management on a large scale.

Organic Management for the Professional

CHAPTER 1

Introduction to Organics

Chemical agriculture specialists have always told us that it makes no difference if you use natural organic or synthetic fertilizer because the microbes have to process the fertilizer into the ion form before plants can properly use it anyway. They are wrong. They forgot an important law of physics: "Nothing starts in motion without a source of energy." All of our energy—food, gas, coal, diesel, etc.—comes from the sun. Only green plants can collect and store the sun's energy and supply it to us in a usable form.

Soil life must have energy. Synthetic fertilizers, for the most part, contain no carbon and no energy. As a result, soil life is forced to use the soil's energy reserves. The soil's energy supply soon runs low, the chemicals don't get properly processed, and plants are fed unnaturally, causing them to have a dramatically lower resistance to pests. As a result, diseases and insect pests are invited in. Production falls off, more chemicals are used, more soil energy is burned up, the soil life doesn't have ample energy to process the fertilizers or detoxify the pesticides, microbe and earthworm populations decline, the soil dies and loses structure, wind and rain erosion take their toll, and the once-productive farm or ranch land soon becomes a biological desert with little diversity and poor health.

Earth is about 30 percent dry land, and only about 11 percent is suitable for food crop production. All life on Earth is supported by a thin layer of topsoil. Fertile topsoil is the result of decaying organic matter, which adds carbon to the soil, and decaying rock. Both must be present. The quality of that thin soil layer determines the quality of the air we breathe, the water we drink, and the food we eat.

Carbon stored in topsoil is the bankroll of plant productivity. The great early civilizations of Mesopotamia, for example, arose because of the richness of their soils and collapsed because of declines in soil quality. Poor land management and excessive irrigation caused soils to become increasingly

Organic programs can be successful at any scale and are better than the toxic chemical approach in every way—including the economics of the project. Photograph by Howard Garrett.

degraded and unable to support the Fertile Crescent civilizations. Ancient Greece suffered a similar fate. The philosopher Plato, writing his *Critias* around 360 B.C., attributed the demise of Greek dominance to soil degradation: "In earlier days Attica yielded far more abundant produce. . . . In comparison of what then was, there are remaining only the bones of the wasted body, . . . all the richer and softer parts of the soil having fallen away, and the mere skeleton of the land being left." What Plato likely did not recognize is how much organic matter had washed away from these Greek soils.

In the New World, similar processes were unfolding. Sylvanus G. Morley, a Harvard professor, concluded that the great Maya civilization of Mesoamerica collapsed because it overshot the carrying capacity of the land. Deforestation and erosion exhausted their resource base, causing many Maya to die of starvation and thirst en masse and others to flee once-great cities, leaving them as silent warnings for generations to come.

According to UCLA professor Jared Diamond, author of the books *Guns, Germs, and Steel* and *Collapse*, most inhabitants of Easter Island in the Pacific died because of deforestation, erosion, and soil depletion. In Iceland, farming and human activities caused about 50 percent of the soil to end up in the sea, explains Diamond. He also notes that Icelandic society survived only through drastically lower standards of living. Not surprisingly, the practice of destroying soils by torching watersheds or salting farms and fields has been employed by armies in warfare from the time of Alexander the Great to Napoleon.

Today, we are facing many of the same issues as these former civilizations: removal of native vegetation, overharvesting, dwindling supplies of freshwater, overworked soils, and sprawling population growth. Our poor management of the land has resulted in serious warning signs. Widespread agricultural pollution of lands and seas, accelerated topsoil loss, damage to fish and aquatic life, pesticide buildup in our own bodies, and the rapidly declining nutritional value of food have become environmental problems of immense importance that are directly related to soil.

Today, the organic content of most agricultural soils is far below sustainable levels. Many soils are down to less than 1 percent, some as low as .2 percent. Originally productive soil organic levels were between 3 and 5 percent and as high as 8 percent by weight. Before World War II and the invention of synthetic fertilizers, no irrigation, fertilizers, or pesticides were needed. Now, the average vegetable farmer spends over $1,000 per acre per year on these three inputs, and the soil quality is still deteriorating.

Sunlight on bare soil, overuse of high-analysis N-P-K (nitrogen-phosphorus-potassium) fertilizers, and overtillage of soils have caused the microbes to deplete the organic matter and humus reserves of the soil. Conversely, natural organic fertilizers, composted manures, and natural organic materials have low amounts of N-P-K but contain a broad spectrum of trace minerals, carbon, and beneficial organisms plus an abundance of energy needed to sustain life.

Plant life collects carbon dioxide from the air, releases oxygen into the air, then loads (combines) the carbon with other elements and energy from the sun to create carbohydrates. This carbohydrate energy the plants manufacture is the only way the sun's energy can be utilized by animals and people in the form of food. This energy is needed by all life-forms for growth and survival.

This sun-generated energy enters the belowground food web as a wide variety of organic substances (i.e., molecules that contain carbon, hydrogen, and oxygen). In the food web, plant residues—leaves, roots, stems, fruits (i.e., soil organic matter)—or sugar exudates from the roots are consumed by the micro- and macro-organisms such as bacteria, fungi, microarthropods, protozoa, nematodes, earthworms, and insects. As carbon flows throughout the food web, it provides energy and the building blocks for the organisms that are consuming it, whose waste or bodies provide the energy and building blocks to other organisms in the web.

In other words, although each organism is working to find energy and food for its own survival, the interactions of all these organisms are providing food and protection to the plants and all the other soil organisms. Ultimately, this organic matter contains the nutrients that new plants need to grow. Then, mycorrhizal fungi in the soil, made up of hyphae that look like fine threads, come in and "hunt" down these nutrients and carry them back to the plant roots.

Research is revealing that practices like reduced tillage, use of cover crops, and incorporation of crop residues can dramatically alter the carbon storage of agricultural lands. Much of the excess CO^2 in the air said to cause global warming comes from the bare cultivated soil on our farm and ranch lands. Parks, golf courses, and landscapes that are managed conventionally also contribute to this problem. For years, many have argued that organically produced food is safer and more nutritious. Now we are learning that a switch to organic production methods is an expedient and soil-based sink for carbon from the atmosphere. Data from the Rodale Institute's long-term comparison of organic and conventional farming methods substantiates that organic practices are much more effective at removing carbon dioxide, the greenhouse gas, from the atmosphere and fixing it as beneficial organic matter in the soil. Organic practices result in rapid carbon buildup in the soil.

In twenty-three years of continuous recordkeeping, the Rodale's Institute's Farming Systems Trial found that organic farming systems have shown an increase of 15–28 percent in soil carbon, whereas the conventional farming system has shown no statistically significant increase. If just ten thousand medium-sized farms in the United States converted to organic production, they would store so much carbon in the soil that it would be equivalent to taking 1,174,400 cars off the road or reducing car miles driven by 14.62 billion miles.

According to USDA studies, a block of soil containing 4–5 percent organic matter (by weight), weighing 100 pounds, and measuring 3 feet by 1 foot by 6 inches deep can hold 165–195 pounds of water. Agricultural soil

Little-known Facts That Have a Big Effect

On a global scale, soils hold more than twice as much carbon, an estimated 1.74 trillion U.S. tons, as terrestrial vegetation does, 672 billion U.S. tons. U.S. agriculture currently releases 1.5 trillion pounds of CO^2 annually into the atmosphere. Converting all U.S. agricultural lands to organic production would eliminate agriculture's massive emission problem. In addition, because fossil fuel–derived chemical fertilizers would be eliminated, switching to organic production would actually result in a net increase in soil carbon of 734 billion pounds.

Organics in a Snapshot

1. Organic matter and its effect on the health of the soil is crucial to the health of all life.
2. Products and techniques should be used that positively affect the health and growth of soil biology.
3. Converting the sun's energy into biological activity in the most efficient manner is the goal.

with this high organic content can absorb 4–6 inches of rain per hour. Most of our agricultural soils today can only absorb .6–1 inch of rain per hour. Water pollution, drought, floods, and soil loss from wind and water erosion are caused, to a large degree, by a lack of organic matter that creates good soil structure.

As organic farmers, ranchers, horticulturists, contractors, and gardeners, we can and should make a difference. True science and common sense support us. The organic program works better in every way. Why aren't more people using organics? The answer is: change. Change of any kind is difficult. The status quo is always very powerful, and often those in charge of it don't want change. Some practitioners don't want the trouble of change, and it is more profitable to vendors of toxic chemicals to treat the symptoms rather than prevent or cure them. There is, however, growing appreciation of the true costs of our synthetic management of the land, and a groundswell of large-scale organic management is sprouting all over America.

Reasons to Go Organic

There are many reasons to go organic and stop using synthetic toxic pesticides and high-nitrogen artificial salt fertilizers. Here are the nine most important reasons:

1. **Improved Health.** Natural organic fertilizers, soil amendments, and pest control products improve soil health and the overall production of plants. They are nontoxic to beneficial insects, birds, lizards, frogs, toads, earthworms, and other valuable life-forms. Pets and humans can be added to the list. Even the manufacturers of the toxic chemicals admit, indeed brag about, the fact that their products kill all the bugs. That's right—they can't tell the difference between the good and bad. Only 1–2 percent of insects are troublesome. Even most disease pathogens are beneficial if they occur in their proper proportions.

2. **Cost-Effectiveness.** Any project can be converted from toxics to organics at the same budget. What is more important is that the organic program costs are reduced every year. Each year fewer sprayings for pests are needed, less plant loss is experienced, and plant yields continually increase. And what price do you put on the health of your animals and family? Garden beds never have to be removed and redone, as is often the case under chemical programs. Under an organic program, the soil gets better and better forever—just as the natural forest and prairie soils do. Organic grass–fed cattle ranches spend little to nothing on fertilizers and pesticides. They rely primarily on the natural recycling of organic matter and the natural control of pests and diseases.

3. **Time Savings.** Because of the healthy soil, irrigation, fertilizing, and replacing diseased and dead plants are needed less often, which saves time compared to the recommended synthetic schedule. Another time convenience is that the timing of the organic applications is not as critical. The season's first fertilization can be applied whenever time and weather allow after the first of the year. The second application may be applied whenever time allows in late spring to early summer. The third application should be done whenever possible from late summer through fall. Since the natural organic fertilizers aren't soluble, they can be broadcast basically anytime. The nutrients become available to the plant roots when the temperature and moisture conditions are right for microbial activity and plant growth. Organic fertilizers stay in place, don't leach through the soil into the water stream, don't volatilize into the air, and basically behave themselves.

4. **Fewer Sick and Pest-Infested Plants.** When adapted plants are chosen and planted correctly in the proper environment, they have few pests and require little care. High-nitrogen synthetic salt fertilizers and toxic pesticides are problem creators. They lead to more pest problems because of the destruction of beneficial microbes and insects. Natural organic fertilizers and organic pest control techniques are problem solvers. They concentrate on stimulating life instead of killing it.

5. **More Stress Tolerance.** Plants growing in soil that is rich in compost, rock minerals, and beneficial living organisms have a greater resistance to all stresses—heat, cold, too much water, too little water, and unusual weather fluctuations. Trace minerals, which are abundant in an organic program, are an important part of this stress tolerance. Organically grown plants will have larger and more efficient root systems. When the proper balance of organic matter, mineral nutrients, air, and living organisms is present in the soil, ornamental plants and food crops have greater concentrations of complex carbohydrates, and these sugars function like natural antifreeze. Anecdotal evidence shows that freeze damage occurs on organically grown plants at sig-

nificantly lower temperatures than on plants artificially grown. Many organic growers have personally experienced these benefits of the natural way.

6. **Better Taste and Food Quality**. The increased trace minerals and complex sugars in plants improve the taste of food crops, which is directly related to the health and nutrition of the plants. So-called experts will argue this point, as they will most of our points, but they are wrong. There's a very easy test: Eat fruits and vegetables grown both ways and see which you like better. Also, the use of toxic chemicals eliminates—or should eliminate—the eating of many vegetables, fruits, and herbs. The best example is the rose. Unless you are spraying poisons on the plants and dousing the soil with toxic synthetics, the rose petals are delicious in salads and teas, and the hips, which are the rose fruits, are delicious in teas and an excellent source of natural vitamin C.

7. **Improved Environment**. Unlike synthetic fertilization, organic products don't cause excess nitrogen to volatilize into the atmosphere or soluble nutrients to leach through the soil and into the water stream. Natural organic fertilizers stimulate beneficial microorganisms, which very effectively clean up contaminations such as pesticides, excess mineral salts, and heavy metals in the soil. Leaching is reduced to almost nil, and no one has to breathe the fumes or otherwise come in contact with the toxic materials. Carbon dioxide, a problem greenhouse gas related to global climate, is taken from the air and converted to an energy form in the soil, where it promotes soil life.

8. **Recycled Valuable Natural Resources**. All once-living materials are rotted through composting and mulching and then recycled back into the soil to build humus and trace minerals. Grass clippings are left on the turf, leaves are left as mulch in beds, tree trimmings are turned into mulch, and animal manures are composted to become important fertilizers. Under the synthetic programs, these valuable resources are often taken to landfills, burned, or otherwise wasted.

9. **Natural Organic Programs Are Fun.** It's really no fun to spill toxic chemicals all over you; breathe the fumes; or have to deal with the contaminated clothes, containers, and residues. It's certainly no fun to worry about developing cancer. Under an organic program, the ease of application, the great diversity of life that's experienced, and the successful production of plants truly create some of life's great pleasures.

Converting to the Natural Organic Program

The first step in converting to an organic program is to stop doing those things that we know to be harmful. These practices should be followed:

Eliminate Bare Soil

Nature never allows bare exposed soil in productive land. Vegetative cover supplies roots and soil with life-giving energy from the sun, whereas bare soil allows wind and water erosion, energy loss, and nutrient depletion. Topsoil loss prevents proper water absorption and good water management, resulting in crop failure and ultimately desertification. Keep the soil covered—by cover-cropping, practicing low- or no-till farming, not overgrazing, and mulching when possible. *Do not allow bare soil to exist.*

Reduce or Eliminate Tillage

Tilling causes the carbon (organic matter) to be oxidized (used up) faster than nature can replace it. Tillage exposes more surface area of soil to sunlight, which kills beneficial fungi, bacteria, and other microbes that store nutrients, help create soil structure, and control soil pathogens. The result is lower yields that are poorer in quality.

Stop Managing Plants Artificially

Artificial salt fertilizers and toxic pesticides are still used because of university recommendations, peer pressure, chemical sales pressure and the related money, paradigm problems, a quick-fix mentality, and laziness. Artificial and toxic products keep unhealthy plants and animals alive and reduce the populations of beneficial organisms in the soil.

The critics will continue to argue that the organic approach is a fad, can't be done on a large scale, and costs too much, but they are wrong. Here's the shorthand version of the ideal program:

1. Stop using all synthetic pesticides and other toxic chemicals. This one is simple. For those faint of heart, cut the current fertilizer applications in half and spend the money saved on the recommended organic inputs. Quitting everything and changing over cold turkey is the best route. Phasing in the organic program is possible, but success will be much slower to obtain. Continuing any toxic products or bad techniques will suppress the desired biological activity.
2. Build soil health with natural organic products (compost and other organic fertilizers, rock minerals, and sugars) and techniques. The best product to add to the soil is quality compost, but because of budget limitations, shortcuts often need to be used. The least expensive way to begin is to stimulate the microbes by adding sugar

or a product that converts quickly to sugar. At the top of this list is molasses. Although any sugar will work, molasses has more trace minerals and does a better job of spiking the growth of the beneficial microorganisms. The sugar primarily stimulates bacteria, but the general increase in life seems to have a balancing effect because fungi populations also increase. The reason dry molasses works better in this regard is that the sugar is best for bacterial growth and the organic piece inside the covering of molasses is best for fungi.

Dry molasses can be used on smaller projects at as much as 20 lbs. per 1,000 sq. ft. For farms and ranches, the rates can be reduced to as little as 50–100 lbs. per acre. Obviously, the more the budget allows, the faster and more significant the establishment of microbes will be. Compost, compost tea, and organic fertilizers that contain mycorrhizal fungi will work even better, but budget again is an issue.

Building a healthy soil is done by putting the basic elements in place and letting nature do the rest. Most unhealthy soils lack air and humus, have a weak population of microorganisms, and are chemically unbalanced. All these things are related, and improving any one of them indirectly improves the others.

3. Use native plants and well-adapted introductions. Help with choosing and planting native and adapted plants should come from regional or even local experts. With that advice as a good starting point, use your own experiments and trials to improve the palette of plants that will best grow on your property, since every property has different conditions. Continue to look at new varieties and even different species of plants to find which ones are best suited to your soil and climate.
4. In cases of extremely poor soil, the first step in major soil improvement is to aerate the ground. Cultivated and pasture land can be ripped or chisel plowed, and turf areas can be mechanically aerated. Landscape beds can be hand-aerated with a turning fork, rototilled, or stirred with tools like the Air Spade or Air Knife.

Basic Guidelines for Stimulating Biological Activity (Life) in the Soil

- Most annuals, vegetables, and grasses do best in bacteria-dominated soils.
- Most trees, shrubs, and perennials do best in fungi-dominated soils.
- Dry, brown, aged, coarse-textured organic material supports fungi.
- Wet, green, fresh, fine-textured organic materials support bacteria.
- Mulch on the soil surface primarily supports fungi.

- Mulch and compost worked into the soil tends to support bacteria.
- Sugars help bacteria grow.
- Seaweed, fish, fulvic acid, humic acid, rock phosphate, and vinegars help fungi grow.

General Notes on Plant Soil Life

- Conifers and hardwood trees prefer ectomycorrhizal (cell-surrounding) fungi on their roots.
- Annuals, perennials, vegetables, herbs, grasses, and softwood trees prefer endomycorrhizal fungi (those whose hyphae enter into the cells of the roots).
- Synthetic fertilizers kill off beneficial soil food-web microbes.

CHAPTER 2

Soil Building

Feed the soil and it will feed your plant.

Basic Soil Science

The proper management of soil is basically the same worldwide no matter what the texture, color, mineral makeup, or pH. Soil is a complex living organism that is dynamic, ever-changing, and very complicated. It is a mixture of rock minerals, air, water, and organic material. It supplies plants and animals with nutrients and life, but it only functions properly if it is healthy. This chapter focuses on how to feed the soil and create the fertility required to produce healthy plants—from crops and turf grass to flowers, roses, shrubs, and trees.

Soil is the outermost layer of our planet that all life depends upon. Five tons of topsoil spread over an acre is as thin as a dime. It takes natural processes at least five hundred years to create 1 inch of topsoil. Over seventy thousand types of soils have been identified in the United States. Five to ten tons of animal life can live in an acre of healthy soil.

Healthy, fertile "garden" loam (topsoil) is composed of sand, silt, clay, organic matter (humus, at least 5% by weight), macronutrients, trace minerals, micronutrients, and microorganisms. Soil with 5 percent organic matter can hold almost 200 pounds of water in every 100 pounds of dry soil. A similar soil with only 1 percent organic matter can hold only 30 pounds of water.

Therefore, for every 1 percent increase in organic matter, the same 100 pounds of soil is able to absorb and hold 30–40 more pounds of water.

Organic matter in the top 6 inches of the soil provides:

- Growth-promoting hormones
- Mycorrhizal fungi
- Nematode-destroying fungi
- Nitrogen-fixing microorganisms
- Increased moisture-holding capacity
- Air space, drainage, and aeration

- Sticky substances that bind particles into aggregates
- Buffer against chemicals and high/low soil pH
- Nutrient availability over a long period of time
- Increased cation exchange capacity (CEC)
- Better soil tilth—easy to work and till, wet or dry
- Help in recycling organic waste products
- Erosion prevention

Factors Affecting Nutrient Availability

Many factors affect soils' ability to store and provide nutrients for plants. These factors are covered in detail in any good soil science textbook. A very good reference is *Humic, Fulvic and Microbial Balance: Organic Soil Conditioning* (1993, Umi Research) by William Jackson, PhD. Soil capacity, plant capacity, and environmental factors must all be taken into account.

Soil Capacity

Amount and type of clay—dispersive, shrink/swell, or nonexpansive

Cation Exchange Capacity (CEC)—total exchangeable cations that a soil can absorb, measured in milli-equivalents per 100 grams of soil; cations have positive electrical charges, such as NH_4^+, K^+, Ca^{+2}, Fe^{+2}, etc.

Anion Exchange Capacity (AEC)—total exchangeable anions that a soil can absorb, measured in milli-equivalents per 100 grams of soil; anions have negative electrical charges, such as NO_3^-, PO_4^{-2}, SO_4^{-2}, etc.

Organic matter—humus or green materials, bacteria food or fungal food

Oxides of iron and manganese—may cause hardpan

pH—the acidity or alkalinity of a soil

Redox (REDuction—OXidation) potential—form of the nutrients and how they react chemically

Sand, silt, clay ratios

With the use of organic amendments and techniques, the microbes and life in the soil will naturally tend to balance and correct these factors except in extreme cases.

Plant Capacity

Plant capacity often determines how well a plant will grow in any given soil. For example, a cactus will not grow well in a rich, fertile soil high in organic matter, and conversely, a shade- and moisture-loving plant like impatiens will

Soil—before and after—organic amendments and in the absence of synthetic fertilizer and toxic pesticides. Photograph by John Ferguson.

not do well in full sun growing in sand or gravel. Many hybrids developed to grow on nutrient-depleted soil can no longer absorb nutrients if they are grown in fertile soil. A live oak from the northern states will not perform as well (i.e., withstand hurricanes and salt effects) along the Gulf Coast as a live oak native to the Gulf Coast.

Environmental and Other Factors

Plant growth is also affected by the environment in which it grows. If the irrigation water comes from a well, it might have dissolved minerals such as carbonates in it. Over time, these dissolved minerals precipitate out and form salts that cement the soil particles together, creating tight, compacted soils that are referred to as hardpan. Municipal water systems might have large amounts of chlorine (or chloramines, which are much worse) in them that is used to kill microbes. Frequent irrigation will also kill the microbes in the soil that help plants grow and prevent disease. A common synthetic herbicide is broken down by bacteria, creating an explosion of bacteria in the soil. Many common weeds love soils with lots of bacteria, hence the weed problems become worse as a condition that favors weeds over the desired plants has been created.

Basic Composition of Healthy Soils

Chemically Balanced Soils

Boron	1.5 ppm
Calcium	60–70% of all minerals present
Chlorides	80–120 ppm
Copper	2–3 ppm
Iron	20 ppm
Magnesium	12–15%
Manganese	10–14 ppm (2/3 of the iron level)
Nitrogen	20–45 ppm as nitrate-nitrogen (N-NO_3)
Organic matter	5–8% by weight (25% by volume)
Phosphate	150–200 ppm as P_2O_5—more than 200 ppm can tie up micronutrients
Potassium	3–5%
Sodium	much lower than potassium, less than 2.5%
Sulfate	20–25 ppm
Zinc	5–10 ppm

Sandy soils will require slightly different levels of these elements than clay soils.

Roots and the Soil Food Web

The underground portion of a healthy plant will always have a greater mass than the growth above. The total root surface area can be over one hundred times greater than the surface area of the trunk, stems, shoots, and leaves combined. The roots' function is to anchor the plant and to gather water and minerals for the leaf surface so food can be manufactured for growth and reproduction.

Roots do not work alone. They solicit the help of many life-forms, especially the microbes. To lure these life-forms to their surface (rhizosphere), the roots give off many products. According to the book *Introduction to Soil Microbiology* by Martin Alexander, the roots excrete all of the naturally occurring amino acids plus forty-nine other important exudates, which include twelve organic acids, thirteen carbohydrates, four nucleic acid derivatives, four enzymes, eight growth factors, and eight other beneficial compounds—collectively known as root exudates.

Soil life such as bacteria, fungi, algae, protozoa, earthworms, and beneficial nematodes feed on these exudates. To show their appreciation for all these goodies, the life-forms help the plants by correcting soil pH and by gathering and processing minerals and water from the surrounding soil for the plants. They also protect the roots from pathogens and harmful nematodes.

There is a whole complex of life-forms in the soil that are living, dining, reproducing, working, building, moving, policing, fighting, and dying to help the plants that feed them. However, we must do our share to help the plants create these root exudates by placing them in their preferred soil, location, and environment with adequate moisture and nutrients. Then this complex of root helpers will build an immune system in the plants against diseases and insect pests.

The root exudates and beneficial organisms are always greatest in rich organic soils under an organic mulch or sod cover. In organically poor and exposed soils, the troublesome insects and diseases are always abundant.

The concept of soil building can be daunting, but it is actually simple: Add materials to the soil that build the life in the soil, and do not add materials that are harmful to soil life.

Soil Biology

Healthy soils contain microorganisms such as bacteria, fungi, actinomycetes, algae, protozoa, yeasts, nematodes, and other tiny critters. There are 930 billion microorganisms in each pound of soil under turf, and there are about 70 pounds of them living and working in each 1,000 square feet of root zone. Many of these organisms are very short-lived, so the turnover is rapid. One hundred pounds of dead microorganisms will contain close to 10 pounds of nitrogen, 5 pounds of phosphate, 2 pounds of potassium, ½ pound of calcium oxide, ½ pound of magnesium oxide, and ⅓ pound of sulfate. With 70 pounds of these creatures per 1,000 square feet of root zone, the poundage adds up to enough per acre for excellent crop production.

Microorganisms feed on organic matter in the soil and form gluelike materials called polysaccharides. These complex sugars are what bind or weld individual soil particles together, creating larger aggregates. This process gives soil its wonderful soft, crumbly texture and helps prevent erosion.

According to soil biologists, a *teaspoon* of healthy soil is teeming with approximately:

- 100 million or so individual bacteria
- 50 to 150 meters of fungal threads
- 10,000 to 100,000 protozoa
- 5 to 500 beneficial nematodes

Pine tree seedlings in growth chamber. The seedling on the right was inoculated with mycorrhizal fungi; the one on the left was not. Photograph by Mike Amaranthus.

Potatoes without (left) and with (right) a biologically active root zone. Photograph by Mike Amaranthus.

Microbes and soil life are extremely important to the growth and health of all plants, to the ecosystem, and to the health and quality of life for humans.

We now know that when we are talking about soil health care and management, first we have to think about microbes and soil life. If we take care of the soil life, then the chemistry and physics of the soil tend to correct themselves. If we focus on the chemistry and physics of the soil first, it is sort of like "putting the cart before the horse," and we get the enormous problems we see from weeds and disease, erosion and soil degradation, and environmental pollution. The first step in taking care of the soil life is summed up in the phrase *Primum non nocere,* which is Latin for "First, do no harm."

Researchers at the USDA's Soil Microbial Systems Laboratory have discovered a new protein named *glomalin,* which is produced by mycorrhizal fungi and helps glue soil particles together, making it easier for air and water to move through the soil. The improved circulation also creates a healthier environment for plants and beneficial microorganisms. Tests have shown that no-tilled corn plots were more stable and contained more glomalin than tilled plots. Soils with the well-stabilized structure were far less prone to erosion from wind or water.

Some of the best ways to increase soil life in both diversity and quantity of species are to use the following:

Compost—an excellent soil amendment that also acts as a slow-release fertilizer containing nitrogen, phosphorus, potassium, trace minerals, microelements, humic acids, fulvic acids, and other needed nutrients (vitamins, enzymes, microorganisms, etc.).

Mulch—the best way to add organic material to your soil, and compost is the best mulch for this purpose in many situations.

Humus (Humates)—the dark brown to almost black substance that is the secret key to healthy soil. It is nearly insoluble and contains about 30 percent each of lignin, protein, and complex sugars. It contains 3–5 percent nitrogen and 55–60 percent carbon. Humus is the source of food and energy for microorganism development and is the stage of decomposition that provides food from the soil to plants in the form of slow release nitrogen, phosphorus, and sulfur.

Organic matter—also stores mineral nutrients such as calcium, magnesium, potassium, and many others. It keeps these valuable nutrients from washing out of the soil and makes them more available to the plant.

Synthetic chemical fertilizers damage soils. As soils are being damaged, plants become weaker and actually attract insects, diseases, and weeds. When we spray fungicides, insecticides, and herbicides, we harm the soil. Organic matter in the form of humus is rapidly used up, eliminating the food source for microorganisms that prevent soil disease and help fix nitrogen from the air. This causes the need for larger amounts of nitrogen fertilizer to keep plants growing, and so the cycle continues. As life in the soil declines, soil structure declines, which greatly increases erosion and allows synthetic chemicals to leach into runoff that pollutes our streams, ponds, and groundwater.

Soils with lots of organic matter hold minerals better. A recent study of agricultural soils found that total soil sulfur decreased in all treatments where no organic material was added. However, with the organic amendments, 26–54 percent of the sulfur remained in the soil with a half-life of twenty-four to thirty-eight years.

One of the biggest problems we have with our soils is the loss of trace and micronutrients (i.e., minerals). The growth and health of plants is limited by the least available nutrient. As plants grow, the minerals are used up to form plant tissue, from stalks and leaves to the fruits and vegetables we eat. If the minerals needed for good health are not in the soil, then they are not in our food (plant or animal), and our health suffers. To restore soil health, we must add back mineral dusts (lava, granite, basalt, greensand, etc.) to the soil. The microbes will release the minerals for plants to use.

Paramagnetism

Researchers have recently identified a physical property associated with good soil, the "paramagnetic force," as it is often incorrectly called. In electricity, we have materials (matter) that are conductors and insulators; similarly, in magnetism, we have materials (matter) that are paramagnetic or diamagnetic. Paramagnetism is just a natural property of certain types of matter. It is a form of magnetism that occurs only in the presence of an externally applied magnetic field. A paramagnetic material will move toward (is weakly attracted to) the source of the magnetic field. A diamagnetic material is repelled (will move away from) the source of the magnetic field.

After studying soils from around the world, it was found that the healthiest soils with the best plant growth and highest crop yields have high paramagnetic values, while poor soils with lots of disease and insect pressure have low paramagnetic values. How these low-level energy fields affect plant growth is not fully understood, but the direct correlation with plant growth is confirmed. Volcanic rocks (lava, granite, basalt, etc.) generally have high paramagnetic values, hence mineral dusts from these rocks have an additional value (see *Paramagnetism: Rediscovering Nature's Secret Force of Growth*, by Philip Callahan).

Soil Texture

Soils are made of three basic components: sand, silt, and clay. Sand particles are the largest in size, and clays are the smallest, with silts in between. The ratio of these components determines the type of soil—sandy loam, clay, sugar sand, etc.—as well as the physical properties of the soil (water-holding capacity, aeration, nutrient-holding capacity, texture, etc.).

Sands: These are the largest and coarsest of soil particles. They are well aerated, ideal for root growth, and rapidly draining, but moisture and nutrient retention are poor. They have the lowest total porosity but highest permeability.

Silts: These particles are between clay and sand in size and properties, are important in loam soils, and have better aeration and water infiltration than clays. They are not particularly important by themselves.

Clays: These are the smallest of the soil particles and have high nutrient- and water-retention capability. They also help bond larger soil particles together and fill in pore spaces between the larger particles.

Soils with a high percentage of clay can be difficult to work, as some shrink and swell with water content. They are often slippery when wet, very hard when dry, and prone to having large cracks; they are associated with poor aeration and slow water infiltration. Clay soils have the highest porosity but the poorest permeability (measurement of the ease with which fluids move through the material).

Loams: Theoretically, these types of soils contain equal amounts of clay, silt, and sand, plus some organic matter, making up the ideal garden soil. In reality, most natural soils are some variation of this ratio. For example, a 40 percent sand, 40 percent silt, and 20 percent clay soil is considered a loam. More specifically, if a soil has 10 percent clay, 65 percent sand, and 25 percent silt, it is considered a sandy loam. A silt loam might have 60 percent silt, 20 percent clay, and 20 percent sand. Any soil with over 55 percent clay is considered a clay soil.

The USGS (United States Geological Survey) and NRCS (USDA—National Resources Conservation Service; http://soildatamart.nrcs.usda.gov/) have soil maps available online to help one identify the soil type in one's area. Most local county agricultural extension offices also have information on a region's soil types. Soil types vary greatly around the world, hence it is always useful to understand the base soil types in your area.

Regardless of the type of soil you have, the microbes and life in the soil will bind soil particles together and create aggregates of soil structure that allow aeration and water infiltration. Even heavy, tight pure clay can be turned into good fertile soil with the use of organic techniques.

Nutrients

Clay soils tend to have the most nutrients in them and sandy soils the least. Organic matter (humus) will have more nutrients and hold them better than clays. Microbe bodies living in the soil contain the most nutrients of all.

Using the natural organic approach, nutrients are supplied in many forms. Soil amendments such as organic fertilizers and rock dusts can provide complete plant nutrition. Justus von Liebig's Law of the Minimum states that all plant growth and health is limited by the required nutrient in least supply.

We really do not understand the role of all the elements in plant growth, much less their role in animal and human health. Researchers have shown that plants will live and grow with a relatively small number of the minerals (elements). Even though the plants will grow without some nutrients being present in the soil, if missing, they will not be in the plants either.

For example, the B-12 vitamin regulates the human immune system. The B-12 molecule is built around a cobalt atom. Bacteria use cobalt to create the vitamin molecule, and then it is absorbed into the plant. Animals or humans eat the plant and obtain the vitamin in body tissues. If cobalt is missing from the soil, we do not get B-12 and thus have to take vitamin supplements. This area of soil science needs a lot more research, as many human, animal, and plant diseases are linked to nutrient deficiencies.

Soil Minerals and What They Do

Below is a list of the properties and uses of common minerals in soil.

Arsenic (As)

- Used in the metabolism of carbohydrates in algae and fungi
- Found in affiliation with the amino acid cysteine
- Too much is toxic to animals and humans
- Too little and mammals will die

Sources: greensand, products from the ocean (seaweed, fish emulsion, crab shells, etc.)

Boron (B)

- Recognized as essential to plant health and growth but not fully understood
- Trace nutrient, 2–75 ppm
- Involved with metabolism and carbohydrate transportation
- Required for certain physiological processes such as those in enzyme and coenzyme systems
- Helps plants use nitrogen and phosphorus
- Strong disease fighter (helps plant immune system's resistance to disease)
- Associated with the prevention of many plant problems: cracked stem in celery, internal cork in apples, black heart in beets and turnips, yellowing of alfalfa leaves, etc.
- Associated with the translocation of sugars in plants
- Related to quality and taste of foods
- Regulates flowering and fruiting, cell division, salt absorption, hormone movement, pollen germination, carbohydrate metabolism, water use, nitrogen assimilation, and other aspects of plant growth
- Often found in soils as the insoluble mineral tourmaline (requires action by microbes to release the boron)
- Flavonoid and nucleic acid synthesis
- Too much is toxic to plants; restricts growth; causes sickly green color often mistaken for nitrogen deficiency; and is associated with root deterioration, bud drop, and poor yield
- Relatively rare in nature, hence deficiencies in soil are common
- Deficiency is often associated with the death of the terminal bud; light green coloring; splintering or cracking of tubers; browning of roots in center; lack of flower formation; small, crinkled, or deformed leaves with irregular areas of discoloration; increased insect and fungal damage and stunting in some plant species, though other species seem unaffected

Sources: compost, native mulches, fish, Solubor, borax, boric acid, biosolids

Cadmium (Cd)

- Too much is highly toxic, will damage kidneys and other organs; also toxic to plant tissue

Source: found in the bodies of microorganisms that live in the soil

Calcium (Ca)

- Involved in pH management; balancing hormone and enzyme systems; strengthening cell walls; making plants healthy by strengthening immune systems to resist insect, bacterial, or fungal attack
- Called the prince of nutrients because it is so vital to soil functioning and nutrient uptake by plants
- Most critical in low-humus soils
- Up to 4 percent of a plant's tissue (compounds) contains calcium
- Used in cell wall construction
- Increased calcium content is associated with increased protein content, which is associated with increased vitamin content
- Involved in enzyme production
- Can improve soil texture
- Makes phosphorus and other micronutrients more available
- Aids in the growth of both symbiotic and nonsymbiotic nitrogen-fixing bacteria
- Aids in the growth of beneficial fungi
- Very important for many microorganisms living in the soil
- Important in water absorption
- Proper calcium helps plants form better stems, leaves, and root systems that allow them to efficiently process sunlight for energy, water, carbon dioxide, nitrogen, and mineral nutrients
- Excessive calcium will cause magnesium, phosphate, and other minor element deficiencies, resulting in poor plant health (ripe for insect, fungal, and bacterial attack)
- Deficiency is associated with stunted roots and stress symptoms on newer leaves (they turn yellow then brown, growing tips bend, and stems are weak), including discoloration and distortion. Also, many weeds grow best in calcium-deficient soil, as their role in nature is to send roots deep into the soil and collect Ca atoms and move them into their leaves and stems. When the plant dies and decays, the Ca is released into the topsoil

Sources: compost, native mulches, lime, bonemeal, colloidal phosphate, gypsum, greensand, and marl

Carbon (C; Organic Matter)

- Between 45 and 56 percent of a plant's compounds are structured with carbon, the basic building block for organic materials (the key to life)

- Significant source of a plant's (and soil's) energy
- Availability is governed by the character of the nitrogen source
- Required in soil as an energy source for microbes and other soil life-forms
- Becomes deficient if too much nitrogen is available

Sources: compost, native mulches, leonardite, humate, coal, carbon dioxide

Chlorine (Cl)

- Essential growth element that influences plant growth but is not fully understood
- Stimulates crops
- Required for strong stalks
- Required for disease resistance
- Plays a role in photosynthesis and oxygen generation
- Rarely deficient in soils
- Plants with a deficiency may exhibit chlorosis of younger leaves and increased susceptibility to disease

Sources: compost, treated city water, organic fertilizers, thunderstorms

Chromium (Cr)

- Used in the bodies of microorganisms that live in the soil
- Deficiency associated with adult-onset diabetes, hence important for any plant used as a food source

Sources: compost, some granite sands, liquid fish, seaweed

Cobalt (Co)

- Used in the bodies of microorganisms that live in the soil
- Constituent of vitamin B-12; in mammals, cobalt is essential for hemoglobin formation and for prevention of nerve degeneration
- Needed in legumes for nodule formation and nitrogen conversion/fixation
- Used in protein synthesis
- Seeds started without cobalt will not grow into a viable plant

Sources: compost, liquid fish, seaweed, greensand

Copper (Cu)

- Beginning to be recognized as essential for plant health; too much is toxic
- Stimulates a plant's natural immune system (disease prevention/resistance)
- Required for certain physiological processes such as enzyme and coenzyme systems
- Vitally important to root metabolism
- Helps in formation of strong stalks
- Helps form proteins, amino acids, and many other organic compounds
- Works as a catalyst
- Involved with pollen formation
- Helps prevent chlorosis, rosetting (the formation of rosettes on the leaves due to disease or insects), and dieback
- Important component of vitamin A
- Involved with regulating photosynthesis
- Excess calcium or nitrogen makes copper unavailable
- Involved in oxidation; component of protein and cyanins, which are found in animals and some plants
- Lack of this mineral in humans may contribute to premature gray hair, wrinkles of the skin, crow's-feet around the eyes, anemia; some aneurysms and strokes may be caused by a lack of this element
- Deficiencies occur in peat and muck soils
- Visual signs of deficiencies in plants include leaf chlorosis and distortion, dieback of young shoots, and poor lignification (often seen as wilting, bent shoots, and lodging [plants growing sideways instead of up])

Sources: compost, copper sulfate, Bordeaux mixture, most organic fertilizers, native mulches, greensand

Hydrogen (H)

- 6 percent of a plant's compounds involve hydrogen (third most plentiful)
- Involved in metabolic activities

Sources: water (H_2O), most organic compounds, hydrogen peroxide (H_2O_2), compost, and native mulches

Iodine (I)

- Influences plant growth but not fully understood
- Important for mammals
- Helps prevent diseases like goiter, hypothyroidism, etc.
- Deficiency in humans leads to usage of fluorine, which disrupts metabolic systems

Sources: compost, liquid fish, seaweed, mineral dusts, some humates

Iron (Fe)

- Considered a micronutrient
- Photosynthesis: required for the production of chlorophyll molecules, although not contained in them
- Involved with many plant processes
- Nitrogen fixation
- Part of cellular enzymes vital to metabolic processes of respiration and energy transfer
- Required for production of certain proteins
- Required for the fixation of nitrogen (part of nitrogenase enzymes)
- Occurs in concentrations of 10–2,000 ppm in plant tissue
- Required for certain physiological processes such as enzyme and coenzyme systems
- Required as a carrier of oxygen in the process of biologic oxidation
- Aids in the prevention of chlorosis
- Best absorbed by plants when it is chelated in an amino acid form
- Unavailable if excess calcium is present
- Plants use iron in ferrous form (Fe^{++}) rather than ferric form (Fe^{+++})
- Solubility decreases about 1,000 times for each whole number rise in pH, thus is more available in acidic soils
- Too much lime (Ca) will induce a form of iron deficiency called lime-induced chlorosis
- Cool temperatures or dry soils can reduce iron availability and cause temporary shortages
- Plenty of fresh organic matter in the soil ensures the availability of iron via the soil-root-microbe system
- Too much phosphate, copper, or manganese can cause iron deficiencies
- Deficiency results in poor uptake of other nutrients, resulting in mineral imbalances in plant tissues; young leaves and areas between

veins on older leaves are yellow, beginning from top to bottom, while veins, margins, and tips stay green

Sources: compost, native mulches, most organic fertilizers, copperas (iron sulfate), rust (iron oxide), chelated iron, greensand

Lead (Pb)

- Too much is toxic to animals and humans; causes intellectual impairment

Sources: some synthetic fertilizers, iron supplements, and mine tailings

Magnesium (Mg)

- Involved in pH management, balancing hormone and enzyme systems, making plants healthy (strengthens immune systems to resist insect, bacterial, or fungal attack)
- Critical in seed germination
- Required for the construction of chlorophyll; each molecule is centered around an atom of magnesium
- Aids in respiration
- Contained in .05–1 percent of a plant's tissue
- Activator of many enzyme systems
- Affects soil pH more than calcium
- Aids in the formation of plant fats, oils, and starches
- Important in formation of fruits and nuts
- Too much will cause phosphorus, potassium, and nitrogen deficiencies by mineral tie up; can combine with aluminum to form toxic substances in the soil that can enter the food chain and cause health problems; causes the soil to crust; reduces aeration; releases soil nitrogen into atmosphere; promotes anaerobic decay, which forms alcohols in the soil
- Tends to be lacking in weathered soils with low pH or not available in soils with excess calcium and potassium
- Deficiency will cause diminished sweetness in crops and fruits; leaf stunting, curling, thinning, and yellowing between veins

Sources: compost, native mulches, Sul-Po-Mag mineral supplement (22% S, 22% K, 11% Mg), Epsom salts (magnesium sulfate or hydrated $MgSo_4$: 10% Mg, 13% S), plant residues

Note: Epsom salts ($MgSO_4$) have been associated with extra growth in peppers and decreased black spot in roses; they are highly soluble and thus act quickly.

Manganese (Mn)

- Micronutrient; plant tissue contains 5–500 ppm
- Required for certain physiological processes such as enzyme and co-enzyme systems
- Enters into oxidation and reduction reactions needed in carbohydrate metabolism
- Needed in seed formation
- Involved in electron transport during the production of chlorophyll
- Involved with detoxification of free radicals
- Indirectly related to NO_3 reduction
- Excess sodium and potassium (10% or more of available nutrients) will cause manganese tie-up, making the manganese unavailable for the plant
- Deficiency results in poor uptake of other nutrients, resulting in mineral imbalances in plant tissues; symptoms frequently resemble those of iron deficiency

Sources: compost, native mulches, manganese sulfates, chelates, weathering of igneous rocks, greensand

Mercury (Hg)

- Too much is toxic to animals and humans; excess is linked to Mad Hatter's disease

Sources: found in some synthetic fertilizers and other soil amendments made from mine tailings or hazardous waste

Molybdenum (Mo)

- Function and requirements not fully understood by scientists
- A trace nutrient; plants require 0.01–10 ppm
- Required by certain beneficial microorganisms that are involved with nutrient uptake
- Used in electron transfer reactions
- Related to other nutrients and their chemical complexing
- Involved with NO_3 reduction
- Required for certain physiological processes such as enzyme and co-enzyme systems (i.e., nitrate reductase and sulfite oxidase)
- Involved with nitrogen fixation, as it is required by some algae, azotobacteria, and nonsymbiotic nitrogen-fixing bacteria
- Helps plants use nitrogen; a shortage leads to inefficient utilization

of nitrogen, leading to increased risk of NO_3 leaching and polluting groundwater
- Used in reduction of nitrates for the formation of proteins
- Critical for grasses and other crops requiring little potash
- Too much can be toxic to plants and animals

Sources: compost, native mulches, most organic fertilizers, biosolids

Nickel (Ni)

- Used in enzyme urease
- Possible use in hydrogenise and movement of nitrogen
- Too much is toxic to plants and animals

Sources: mineral dusts, liquid fish, seaweed

Nitrogen (N)

- Used in 3 percent of a plant's compounds
- Important for amino acid and protein production; accounts for 16–18 percent of a plant's proteins
- Required for leaf growth
- Most useful forms of nitrogen are found in organic matter and released through microbial activity
- Texas soils require less nitrogen than soils anywhere else in the United States
- Synthetic nitrogen can delay maturity of many plant species
- Chemical form of nitrogen supplied to plants affects protein synthesis; results in lower protein content than from natural sources; contributes to much higher rates of disease; and attracts pest insects
- Most misused of all the fertilizer elements
- Excess creates an imbalance that hurts plant growth, reduces sugar content, and delays crop maturity. Too much nitrogen creates weak succulent growth that attracts insects and pathogens and causes fruits to crack, bitter pits, tip burn of leafy vegetables, tomato blossom end rot, browning of cauliflower curds, new-growth dieback, and drooping flowers on roses
- Nitrogen as NO_3 leaches very easily, polluting groundwater and the air we breathe

Sources: compost, organic fertilizers, rain, air, microbes, blood, and animal tissue

Oxygen (O)

Oxygen is an often overlooked but extremely important element when dealing with soils and plants.

- Healthy soil is 25 percent air (oxygen is the critical ingredient of air)
- 43–45 percent of all the compounds in a plant contain oxygen
- Adding oxygen to soils often creates an immediate growth response; a lack of oxygen creates conditions in which diseases and pests thrive and weed seeds are encouraged to germinate
- Critical for all beneficial biological processes in the soil

Sources: compost, water (H_2O), air (O_2), hydrogen peroxide

Phosphorus (P)

- Concentrates in seeds and fruits
- Contained in 1 percent of all plant compounds
- Involved with stimulating many plant activities (called a "go" food for plants)
- All plant tissues contain phosphorus
- Directly related to root and fruit development
- Required for sugar development
- Aids in disease resistance
- Mineral forms quickly become insoluble and unavailable to plants, whereas organic forms are readily absorbed and used by plants (another value of compost)
- Requires soils with high organic matter content for the microorganisms to make phosphorus available to plants (microorganisms excrete acids that release phosphorus from mineral forms). *Note:* Humic and fulmic acids in compost help release phosphorus
- Part of the molecules that make up genes (DNA and RNA)
- Plays a role in the molecules used for energy storage and transfer
- Sometimes called the soil's catalyst
- Required for plants to use nitrogen
- Deficiency shows up in plants as stunting; slow growth; reddish or purplish cast to leaves; tips dying off; delayed maturity; seeds, tubers, and grains suffering or failing to develop

Sources: compost, native mulches, colloidal and rock phosphate, phosphoric acid, manures, biosolids, bonemeal, fish and seaweed by-products. Soft phosphate rock is best slow-release source of this nutrient

Note: Researchers at Pennsylvania State University have found that excess phosphorus is detrimental to plants in that it decreases drought tolerance and stress resistance in general.

Potassium (K)

- Involved in pH management, balancing hormone and enzyme systems, making plants healthy (strengthens immune systems to resist insect, bacterial, or fungal attack)
- Contained in 3–6 percent of a plant's compounds
- Regulates water movement within plants
- Regulates the balance between root and leaf growth
- Involved in translocation of vital sugars in plant structures
- Strengthens plant stalks and thickens plant tissue
- Helps undo stress caused by too much nitrogen in the wrong form
- Serves as a catalyst for many processes
- Required in the building of chlorophyll
- Required for plants to absorb elements out of the air, such as carbon, hydrogen, and oxygen
- Required for the production of starches, sugars, proteins, vitamins, enzymes, and cellulose
- Aids plants in surviving drought conditions
- Aids plants in increasing both winter (cold) and summer (heat) hardiness
- Governs resistance to certain diseases
- Associated with decreased damage from insects
- Required for the efficient utilization of all nutrients in the soil
- Aids in root growth
- Involved in balancing nitrogen and phosphorus
- Required in soil by microbes to fix nitrogen into the soil from the atmosphere
- Generally occurs in sufficient quantities in most soils but often unavailable due to mineral imbalance
- Excess may cause magnesium imbalance
- Deficiency contributes to winter kill, poor survival of perennials, increased susceptibility to diseases. Other symptoms include young plants with dark green leaves that have small stems and short internodes or are wrinkled and curled; older leaves that are scorched on margins; weak stems, shriveled fruit, or uneven ripening

Sources: compost, native mulches, granite dust, greensand, potassium sulfate, molasses, potash

Selenium (Se)

- Used in the enzyme for glycine reductase found in some cells
- Important in protecting humans against chronic degenerative diseases, as it is required in the production of powerful antioxidants such as vitamin E and glutathione peroxidase. America's "Stroke Belt" encompasses an area where selenium content in soils is low.
- The amounts of beta-carotene and vitamins C and E contained in herbs (mints) are linked to the amount of selenium in soil. The effectiveness of antioxidants in our bodies has also been linked to the presence of this mineral.
- Helps eliminate toxic heavy metals from the body
- Colon cancer survivors with highest levels of selenium were least likely to have a reoccurrence
- Too much is toxic to animals and humans
- A deficiency of this mineral may lead to cardiomyopathy, joint problems, muscular dystrophy, cancer, heart disease, cirrhosis of the liver, and cataracts

Sources: Brazil nuts, free-range meats such as turkey and pork, and compost containing these things

Silicon (Si)

- Influences plant growth but not fully understood
- Used in plants' structure
- Beneficial effects include enhanced resistance to disease and environmental stress, improved growth and quality, photosynthesis stimulation
- Researchers at the University of Florida have found that silicon boosts the disease resistance of rice crops to fungal diseases; University of Oklahoma researchers have found that silicon increased stem thickness, height, and size of flowers; Japanese researchers found that silicon prevents stem blight on asparagus

Sources: diatomaceous earth, zeolite, algae and bacteria, sand, mineral silicates, and most volcanic rock

Note: Best applied in small amounts to foliage or soils.

Sodium (Na)

- Involved in pH management, balancing hormone and enzyme systems, making plants healthy (strengthens immune systems to resist insect, bacterial, or fungal attack)
- Twelfth most common element found in plant tissue
- Perhaps most important to plants in association with potassium (available K must exceed available Na or problems will occur)
- Influences plant growth but not fully understood
- Too much will stunt growth or kill most plants
- Rarely deficient in Texas soils

Sources: compost, most manures, wood ashes, greensand

Sulfur (S)

- Contained in .05–1 percent of a plant's tissue
- Now considered a primary plant nutrient; used by plants in the sulfate form
- Required in legumes (nitrogen fixers) for good nodule development
- Easy to mistake a sulfur deficiency as a nitrogen, magnesium, iron, or available potassium deficiency
- Plants use a lot of sulfur
- All organisms use sulfur in the form of amino acids (cysteine and methionine) that are used to build proteins
- Sulfur is a component of vitamins, coenzymes, thiamine (vitamin B1), biotin, and lipoic acid and participates in many enzymatic reactions
- Soil microbes are responsible for converting sulfur into a sulfate form that plants can use
- Required in the production of proteins and seeds
- Improves the taste of foods
- Easiest leached of all mineral nutrients
- Forage crops grown on sulfur-deficient soils have lower protein content and higher nitrate concentrations
- Deficiency is signaled by sick crops; insect, bacterial, and fungal attack; upper leaves turning yellow, stems staying small and woody, roots becoming long and slender

Sources: compost, native mulches, molasses, sulfates, gypsum, elemental sulfur

Zinc (Zn)

- Required for healthy plants, involved in plant immune system
- A micronutrient; plant tissue contains 3–100 ppm
- Weed populations tend to increase in soils deficient in zinc
- Required for certain physiological processes such as enzyme and co-enzyme systems. Includes alcohol dehydrogenase, superoxide dismutase, carbonic anhydrase, RNA polymerase
- Used as an enzyme activator
- Used in carbohydrate and protein metabolism
- Required by azotobacteria and other nonsymbiotic nitrogen-fixing bacteria
- May act in the formation of chlorophyll
- Aids in the prevention of chlorosis in some plants
- Stimulates plant growth
- Aids in bud development
- Prevents the occurrence of mottled leaf in citrus, white bud in corn, and other disorders
- Aids in the formation of enzymes and hormones
- Used in anhydrases, dehydrogenases, proteinases, and peptidases
- Important for the sweet taste in vegetables and fruits
- Availability related to soil drainage (well-drained soils increase availability)
- Deficiency: terminal leaves are small, bud formation is poor, leaves have dead areas, yellow or mottled intervein regions

Sources: compost, native mulches, kelp meal, seaweed and its extracts, zinc sulfate

Growing evidence indicates that trace and micronutrients are essential for proper plant, animal, and human health. Some soils have plenty of these, but others have very few or none. If the soil has abundant organic matter and a healthy functioning microbial system, then the microbes will help the plant obtain and absorb the ones they need in the correct amounts.

Trace mineral tests have been conducted on cole crops (Brussels sprouts, cauliflower, cabbage, broccoli) in a fertile compost-amended soil. Every other transplant had a tablespoon of trace minerals added to the planting hole. Only the Brussels sprouts had a huge growth and heat-tolerance response. The tests indicated that the Brussels sprouts' health, growth, and production were limited by a trace mineral deficiency.

Trace and micronutrients are found in products from the ocean (fish emulsion, seaweed), greensand, and granite fines as well as other products (see "Soil Amendments" below). The use of these minerals is a cheap way to ensure healthy plants, especially if the latter are to be used for animal or human consumption.

One of the roles played by plant species called weeds is to mine these nutrients from deep in the subsoil and bring them up to the surface in their aboveground parts. When the weeds die, microbes decompose the plant material, releasing the nutrients into the topsoil for the next generation of plants to use. For example, the dandelion plant with its long taproot will collect calcium ions from deep in the subsoil. Since the dandelion is an annual weed, each year calcium is collected in the leaf and stem tissue and then released after the plant dies and decomposes. When enough available calcium has occurred in the correct ratios to other minerals, the dandelion seed will no longer germinate, as it has done its job in nature. Many weed problems are associated with soil health and fertility problems. When these are corrected, the weed problems decrease or disappear.

Soil Testing

Have soil tested, preferably by a lab that gives organic recommendations, to learn the total and available levels of organic matter, nitrogen, calcium, magnesium, phosphorus, potassium, sulfur, sodium, chlorine, boron, iron, manganese, copper, and zinc. Tissue sample tests are important to see what nutrients are being taken up by the plants. The best lab for organic recommendations is Texas Plant and Soil Lab in Edinburg, Texas. They may be reached at 956–383–0739.

Simple Soil Tests

There are some simple tests one can do to determine what basic soil types one has without the cost of a lab test.

1. Take a core sample of soil 6–8 inches deep, place it in a straight-walled jar (e.g., olive jar) that is 75 percent full of water (leave some air space at the top), and shake vigorously for 10–15 minutes until all soil chunks are dissolved. Let the jar sit overnight or longer until the water on top of the sediment is clear. You will see different layers of soil types. On the bottom are the heaviest and largest particles: gravel, then sand, silt, clay, and finally organic matter on top. Measure the total thickness of all sediment layers with a ruler (metric rulers are easier to use). Measure the thickness of each layer, and then

divide the layer thickness by the total thickness to give the amount of each type of soil component present.

2. Get a handful of soil and wet it a little with water. Squeeze your hand into a fist and then release. All but the sandiest soil will form a ball or clump with distinct finger marks. If you can make a rope with the ball by rolling it between your hands, then it is a good mix of sand, silt, and clay. If you can flatten the rope into a ribbon without breaking it, then it has a high clay content.

3. Rub a small amount of soil between thumb and forefinger:

- If it feels gritty, then the soil is sandy.
- If it feels like moist flour or talcum powder, then it is silty.
- If it feels slippery and moldable or sticky, then it contains a lot of clay

Soil Amendments

Basalt: Available in many forms: sand, crushed, weathered, etc. It is an igneous rock formed from molten lava that is low in oxygen but rich in iron and magnesium. It is also paramagnetic, although values vary based on exact type. Basalt improves all soil properties.

Bottom Ash: Produced from the burning of coal but often contains lead, arsenic, mercury, cadmium, and other toxic materials that become concentrated in the ash. It is not recommended for use in horticulture.

Coir: Shredded fiber from coconut husks. Slow to decompose, this renewable resource helps loosen and aerate soils and is used in potting mixes. Contains nitrogen, phosphorus, potassium, and trace minerals and is an excellent product for stimulating biological activity in the soil. Often woven into mats and used to line hanging baskets.

Cornmeal: Cornmeal in any form serves as a soil amendment and mild natural fertilizer. Because it both contains and stimulates a beneficial fungus called tricoderma, it also has fungal disease control properties. The outside edge (bran and germs) of the corn kernel (hominy) has been called horticulture cornmeal. Whole-ground cornmeal is the most commonly used form for soil amending and disease control. Corn gluten meal is the protein fraction of corn and has fertilizer and preemergent weed control ability. Corn gluten feed is a low-quality waste product, and the regular cornmeal sold in grocery stores is basically the starchy endosperm of the kernel and has low value as a food or horticulture product.

Diatomaceous Earth (DE): Occurs naturally and in a calcined form that is used in filter media. Used as a source of silicon. Raw DE has moisture-absorbing ability and is used in insect control. It is also used in animal feeds for internal parasite control and to pull toxins out of the body. The DE used for swimming pools and other filter media is dangerous to breathe and does not have any use in plant or animal management. Always look for food- or agricultural-grade DE.

Epsom Salts: Also known as magnesium sulfate ($MgSO_4$). Provides magnesium and sulfur and does not alter pH. Small amounts help remove sodium from soils.

Expanded Shale: Produced from clay minerals that are found in nature in various forms. The clay minerals or rock is crushed and then fired in an oven at high heat, which causes the minerals to expand into a porous rocklike aggregate that is screened into various sizes. These aggregates are lighter than gravel, are very hard, and are used to make lightweight concrete. They are frequently used as an ingredient in soil mixes for green roofs, as they hold their structure and are lighter than soil and sand.

Fly Ash (FA): Produced from the burning of coal for electric power generation. Tends to be very alkaline but contains many minerals useful as plant nutrients. It works best in acidic soils, helps stabilize minerals in a form most beneficial to plants, speeds up biodegradation of some pesticides and herbicides in soil, and adds valuable micronutrients such as molybdenum (Mo) that are often deficient in soils.

On the negative side, fly ash can cause boron phytotoxic effects in plants, and its high salt content can cause phytotoxic effects in others. Too much reduces nitrogen fixation from the atmosphere by microbes, prevents plants from absorbing essential nutrients, and can raise pH to a level harmful in some plants.

Granite: Available in many forms: sand, crushed, weathered, etc. It is an igneous rock prized by gardeners for its ability to improve plant growth and many soil properties. Granite is composed of oxygen-rich minerals that slowly release many nutrients over time by the action of soil microbes. Granite is also paramagnetic, although values vary based on exact type and source.

Rock minerals and sugars—lava sand on the left and dry molasses on the right—are as important to the organic soil system as compost and organic fertilizers. Photograph by Howard Garrett.

Greensand: A naturally occurring greenish mineral powder with the texture of fine sand. The most common type is found in New Jersey, so it is often called Jersey greensand. It is a primary source of potassium but also contains iron, magnesium, calcium, phosphorus, and up to thirty other trace minerals necessary for plant health in an insoluble form. These minerals must be digested by microbes in the soil before becoming available for plants. Greensand is a proven soil conditioner that enriches and mineralizes most soils. It is the most cost-effective source of plant nutrients, often containing up to 12 percent iron. For a full paper on the benefit of greensand, see: www.natureswayresources.com/factandinfosheet/greensand.

Gypsum ($CaSO_4$): A naturally occurring mineral made up of calcium sulfate and water ($CaSO_4$+$2H_2O$) that is sometimes called hydrous calcium sulfate. It is the mineral calcium sulfate with two water molecules attached. By weight, it is 79 percent calcium sulfate and 21 percent water. Gypsum has 23 percent calcium and 18 percent sulfur, and its solubility is 150 times that of limestone, thus it is a natural source of plant nutrients. Gypsum naturally occurs in sedimentary deposits from ancient seabeds; it is mined and made into many products, like drywall, used in construction, agriculture, and industry. It is also a by-product of many industrial processes. One of the major uses of gypsum is to remove sodium salts from soils. For a full paper on the benefit of gypsum, see: www.naturesway resources.com/factandinfosheet/gypsum.

Humates: The partially decomposed remains of ancient plants that have been buried and exposed to heat and pressure over geological time but not long enough to form coal. Excellent source of humic materials that help soils retain nutrients and provide a food source for many beneficial microbes.

Lava Sand: A by-product of mining and crushing lava to make landscape gravel and rocks. Depending on the minerals in the lava, it may be reddish or black. The magma that produced the lava flows often cools relatively quickly in geologic terms, leaving the material paramagnetic. The crushing and screening process produces very small particles (silt sized to fine sand) that are called lava sand. These smaller particles will help coarser soils hold water better and do not decompose over time, as would humus or compost, so they can permanently change the physical properties of the soil. The beneficial properties of lava sand vary greatly, depending on the source.

Leonardite: Oxidized lignite coal that is composed mainly of humic substances (humic and fulmic acids) that increase the availability of macro- and micronutrients. These substances also promote plant growth and make fertilizers more efficient so less is required.

Lime (CaO): Sometimes called quick lime, as it is very reactive chemically, it is produced by heating limestone to high temperatures. Lime is very alkaline and is used to make mortar, plaster, cement soil, concrete, and other industrial compounds. It is not the best calcium source for horticultural applications.

Limestone ($CaCO_3$): Used as a natural calcium and magnesium source and to raise the pH of the soil. The extra calcium reduces the negative effects of aluminum on plant growth. Most limestone also contains various amounts of magnesium in the form of $MgCO_3$. When the magnesium content is over 10 percent, then it is called dolomitic lime or dolomite, as it contains both $CaCO_3$ and $MgCO_3$. More recently, it is being used as an ingredient for green-roof soil mixes.

Peat Moss: Used for many years as a soil amendment and in potting mixes, it is no longer recommended, since it is not a renewable resource because wetlands have to be destroyed to harvest the peat moss. It does not support beneficial microbes or provide nutrients as would good compost. It is a good product to use for winter storage of perishable materials such as bulbs because of its antimicrobial properties—just the opposite of what the soil needs.

Perlite: Mined from a silicon-rich volcanic rock that has been heated to 1,800°F. The heating causes it to expand into a lightweight porous sterile medium. It has a gritty feel, resembles foam pellets, and often floats in water. It contains no nutrients but does help improve drainage and aeration of some soils; however, it does not break down or deteriorate in the soil.

Pumice: A volcanic glass often formed from volcano ash that has lots of bubbles. It tends to be a very porous, lightweight material and has been used to increase the water-holding capacity of soils.

Sea Minerals: This is a group of products produced from seawater. Seawater has over ninety minerals in it, hence products derived from the ocean often contain the same minerals. Some of these products are just sea salt or minerals produced by removing the water. Others are mined products from ancient sea-salt deposits that have been weathered and much of the sodium leached out. Two of the best products in this category are Redmond Sea Salt and Redmond Clay.

Sul-Po-Mag *(or K-mag or sulfate of potash-magnesia):* A mixture of K_2SO_4 and $MgSO_4$ used to provide these nutrients to soils.

Vermiculite: The soft, shiny stuff found in potting mixes, this aluminosilicate clay mineral is similar to mica. It is mined and heated to 2,500°F, which causes it to expand into sterile lightweight particles. It can hold three to four times its volume in water and can improve aeration and drainage of some soils. It also helps to hold cations in some soils (improves CEC).

Wood Ash: Produced from the burning of wood. It is becoming more available as more wood is being burned for energy production. It is a good source of plant nutrients if used in small amounts at one time. Wood ash is alkaline, so large amounts may affect the pH of the soil. Wood ash has been used for centuries as a source of plant nutrients.

Zeolite *(aluminum silicate):* Derived from ancient volcanic ash made of silicate minerals formed by the interaction of sea salt with molten lava, which creates a very interesting physical structure with lots of cavities and voids. It has the ability to absorb water even under dry conditions, or to absorb huge amounts of gas or minerals and keep them readily available for plants. It is often used as cat litter, as it absorbs odors. When used as an amendment, it increases the CEC (cation exchange capacity) of soil-less media. Over forty-eight types of zeolite with similar properties have been found in the United States. Zeolites have the ability to buffer and balance the chemistry of the soil. They will hold certain elements, such as ammonia, that are in excess and release them as biological activity develops.

CHAPTER 3

Planting

Planting properly is important so that plants have the best chance of staying alive, being healthy, establishing correctly, and growing efficiently. Even though many plants can survive being planted incorrectly, they will never grow as well as they might have to provide beauty, shade, and food. Natural organic techniques work best because they work with nature's laws and systems and they improve the soil instead of degrading it.

Basic Bed Preparation

Improving any kind of soil on any size property for planting seeds or transplants requires a simple formula: (1) space for the new plants, (2) air in the soil, (3) organic matter (humus) and beneficial organisms in the soil, (4) available trace minerals, and (5) water. Ranch land can be improved economically by mechanical aeration and/or ripping and adding a carbon material. The most cost-effective additives are humates and raw sugars. Farms, orchards, and landscapes do better more rapidly with more intensive amendments—compost, manures, rock powders, sugars, biological inoculants, and natural organic fertilizers. Green manure cover crops are good for all large sites.

New landscape beds should be prepared by scraping away existing grass and weeds and putting them in the compost pile and then adding the following amendments:

- 4–6-inch layer of compost
- Lava sand, decomposed granite, or zeolite at 40–80 pounds per 1,000 square feet
- Natural organic fertilizer at 20 pounds per 1,000 square feet
- Dry molasses and/or whole ground cornmeal at 10–20 pounds per 1,000 square feet

Till the amendments into the native soil to a depth of 3 inches. Excavation and additional ingredients such as concrete sand, topsoil, and pine bark are unnecessary and can even cause problems.

Notes: More compost is needed for shrubs and flowers than for groundcover. Add greensand to alkaline soils and powdered high-calcium lime (lime [CaO] or limestone [$CaCO_3$]) to acid soils. A biological inoculant, such as mycorrhizal fungi, can also be incorporated into the soil during preparation for planting.

Plants should be removed from containers and soaked in water with Garrett Juice (for formula, see Appendix 1) and New Plant Thrive (http://www.naturalorganicwarehouse.com) or any quality bacteria and mycorrhizae product, mixed in according to label directions. If all soil is removed from the roots, the roots should be treated with mycorrhizal fungi gel. Plants with sopping wet roots should be planted in moist soil. The crowns of plants should ideally be set slightly higher than ground grade but at least level with the surface.

Planting on a large or commercial scale is not that different from planting the home vegetable garden except that different equipment needs to be considered due to economics. Seed treatment is also beneficial. Seed can be treated with a variety of materials. For example, biological stimulants such as seaweed will increase the germination speed and in most cases promote a higher percentage of germination, but the cost benefit needs to be weighed. The most effective products are those that are micronized and contain mycorrhizal fungi. Treating roots of transplants with mycorrhizal fungal inoculants before installation is highly effective.

Seed Planting

Hand Selecting and Saving Seed

Seeds are the least expensive way to plant. For long-term success, it is best to plant open-pollinated (pollinated by natural means such as insects, birds, or wind) species and start the process of natural selection. Each year, select the best of the early fruit from the very best plants and use that seed for next year's crop. Continue that process year after year, and through this selection of the best from the best, your plants' quality and adaptability will get better each year.

Soaking seeds or just moistening the seeds with compost tea or a product like Garrett Juice helps germination and growth. For many seeds, it takes the microbial activity to help break through the seed shell or covering to get moisture to the seed germ. This soaking also gives the plant a boost in nutrition.

Some commercial seed inoculants on the market are definitely worth their money, not only for helping emergence but also in preventing nematodes and root diseases.

Planting Seeds Outdoors

Seeds should be broadcast at the proper spacing, and the soil should be firmed with the back of a hoe or a board, giving the seeds good soil-seed contact. Seeds can then be covered with a very thin layer of earthworm castings or screened compost.

The seedbed should be watered gently and kept moist until the small seedlings come up. The watering should then be cut back to the proper watering schedule and amount for each particular plant species. For small areas, an effective technique to improve germination is to cover the seeded area with moist burlap, which serves to shade the seedbed and keep it evenly moist. Remove the burlap to save and reuse it after the seeds have sprouted.

Thinning is a controversial subject. Ruth Stout, gardening author (*Gardening Without Work*, 1998) and mulching proponent, says that thinning is unnecessary, as the small plants she leaves always have good production. Many other gardeners believe seedlings remaining after thinning will have plenty of space for the root systems and top to grow. With wide spacing, the roots will go deeper. Thinning can be done by pulling a rake through the seedbed when the seedlings are about an inch tall. It can also be done by hand clipping the seedlings with scissors. These cuttings (if herbs or vegetables) can be used in salads. Crops that can be thinned with a rake include parsley, chard, collards, kale, Chinese cabbage, radishes, lettuce, kohlrabi, spinach, turnips, rutabagas, mustard, beets, and carrots. It is not a good idea to thin beans or peas with a rake because this may damage tender stems. Most crops need only one pass with the rake. Try both methods and see what works best for you. These same techniques work for ornamental plants as well.

Starting Seeds Indoors

Most vegetable seeds need a constant warm temperature, about room temperature, to germinate. Most seeds don't need light to germinate. Floating row cover can be put over the flats or pots to help maintain the constant temperature and moisture level. Some seeds are temperature sensitive and need cooler or warmer soil temperatures to germinate. It is recommended that a soil thermometer be used to verify soil temperatures and ensure success. A good soil thermometer can be purchased for only $8–$15 each. (An excellent book on seed germination is *The New Seed-Starter's Handbook* by Nancy Bubel; see Information Resources.)

Seed flats should be checked daily. After sprouts emerge, remove the covering. To prevent seedlings from leaning toward the light, turn the pots or flats around at least every two days. If the seedlings don't get enough light, they'll stretch and get weak and spindly. Tomatoes and onions are about the only plants that can recover from legginess. Some farmers recommend planting tomatoes deeply and burying most of the leggy stems. That may not be the best idea. Most other plants do not respond well to this deep-planting technique.

Water the seedlings gently. If you cover your flats and trays, the soil will stay moist until you remove the covering. After that, check the soil several times a day with your fingers. If it is dry, add water. Plants on a table near heating devices will dry out quickly and may need water more than once a day. Seedlings are delicate and shouldn't be over- or harshly watered. They can easily collapse under the weight of the water. Water gently around the edge of the container or at the base of each plant.

Bottom watering is preferred by some growers and is used commercially to water thousands of pots of any size at one time. Submerge the flats in water and then let the excess water drain off. This method ensures even coverage and wetting of the soil in the pots. Other growers oppose this method because it tends to keep the soil too wet and pushes out the oxygen. We prefer watering from above. A method John has used to water a lot of flats simultaneously is to lay a rectangle of two-by-fours on a level surface and cover with plastic film. The flats are placed into the basin created, and it is filled with water (or water and seaweed solution), soaking all the flats at once.

Natural organic fertilizer can be added to the soil after seedlings are about an inch tall. Earthworm castings and compost are very gentle and may be the best choices for young plants. Garrett Juice can be added to the watering at 2 ounces per gallon. Even though it is a foliar-feeding mix, it is also an excellent mild fertilizer for the soil and a root stimulator.

Seedlings need to be hardened off for a week or so outdoors before being transplanted into their final place in the garden. Introducing young plants to an outdoor environment should be done gradually on a mild day. Plants should be left in partial shade and protected from wind for a few hours the first day, then given a little more exposure time each day. It normally takes three or four days to accustom young plants started indoors to direct sunlight outdoors. In about a week, the plants can stay out all day in the full sun. Seedlings can be moved back indoors during the transition period if the weather changes abruptly. If all this sounds like a lot of trouble, it is, but it's worth it for your ultimate vegetable or herb production.

Inoculation with Beneficial Soil Organisms

Inoculants are inexpensive and an easy way to add rhizobial bacteria and mycorrhizal fungi to the soil. For rhizobia, be sure to buy the right inoculant for the seed you are planting. The package will list the plants affected by the product. Most states have laws regulating the quality of inoculants on the market, so you can usually trust what you're buying. Cold has no effect on inoculant bacteria, but never store them in the sun or in hot storage areas. Legume bacteria can be added as a dry powder or to the soil at planting. If you're planting a mixture that requires several types of inoculants, they can be mixed together. Plant the seeds immediately after inoculating to avoid reducing the efficiency of the bacteria. These bacteria are completely harmless even if you have open cuts in your skin. Reinoculate any seed kept overnight. Don't use any synthetic fertilizers, pesticides, or other harsh seed treatments. Like nearly all beneficial soil organisms, seeds are sensitive to these foreign toxic substances, but mild organic treatments such as seaweed can be used for added benefit. Humus in the soil provides the ideal living conditions for the nitrogen-fixing bacteria species.

More recently, advancements in our understanding of mycorrhizal fungi and their requirements have led to the production of high-quality mycorrhizal inocula for soil and plants. Mycorrhizal inocula are currently available in granular, powder, seed-coat, and liquid forms.

The most important factor for reintegrating mycorrhizae into planting operations is to get mycorrhizal propagules near the root systems of target plants. Inocula can be incorporated into the planting hole at the time of transplanting, watered into porous soils, mixed into soil mixes, or directly dipped on bare-root systems using gels. For agricultural purposes, it is best banded (applied in bands or side-dressed) or applied with seed at sowing. The form and application of the mycorrhizal inoculum depends on the needs of a given planting situation and the equipment used by the professional. It is clear that on projects where mycorrhizal fungi have been lost, inoculation can greatly aid plant establishment and performance.

Cuttings

Cuttings and tissue culture are the best way to reproduce the exact plants that have been successful. We recommend two books on plant propagation: *Plant Propagation Made Easy* by Alan Toogood and *The New Seed-Starter's Handbook* by Nancy Bubel (see Information Resources). For both softwood and hardwood cuttings, we recommend using an organic potting soil that has had extra volcanic sand or other coarse-textured pieces added for increased drainage. One of the successful tissue culture operations in Texas is Magnolia Farms in Houston; another is Casa Verde in Dallas.

Many herbs, annuals, perennials, vegetables, and even tropical plants can be started from cuttings. Take them when the plant is actively growing for best results. Water the plant a few hours ahead of when you plan to take the cuttings so it will be full of water. Cuttings from young plants may be taken anywhere there are sprouts long enough to be removed. If the plant is old, choose a young sprout. Herbaceous cuttings are usually from 3 to 6 inches long. A smooth, clean cut will form a callus and heal better than a ragged one. Cut the bottom end on an angle to allow more surface area for rooting. Pinch off all flowers, flower buds, forming seeds, and all except two to four small leaves before inserting it in the rooting medium.

Insert the cuttings into a moist, well-drained rooting medium, such as damp granite sand or a combination of granite sand, coconut fiber, and fine-textured compost. A mixture of peat, compost, and volcanic rock powders or sand is also good, but peat moss is expensive and antimicrobial.

A fleshy growth called a callus will form on the cut end of the cutting; this is the healing over of the cambium layer. It often, but not always, precedes rooting.

Most cuttings should be stuck into the medium as soon as they are made so they won't dry out. A blunt pencil or a dibble is a handy instrument for making a hole where the cutting will be inserted. Don't insert the cutting before making a hole first, or you may bend or damage the cut end. Insert the cutting upright so the bottom end is about 1 inch below the surface of the medium. Keep the medium moist but not wet, provide a temperature of 70°F–80°F, and keep the cuttings out of drafts.

Plants with heavy, sticky sap content such as geraniums, pineapples, or cacti are less likely to rot before they root if they are spread out on newspapers in the shade and allowed to dry for a few hours before they are inserted in the rooting medium.

Cuttings can be treated with rooting hormones first, but it's not essential. Homemade rooting hormones can be made from natural apple cider vinegar in water or willow water, which is made by soaking 6–8-inch willow stems in water overnight. Human or animal saliva works even better.

Many plants can be started from cuttings. Most groundcovers, vines, shrubs, and perennials are started this way. Most trees are now started from cuttings to give true genetics to the new plants. As opposed to starting plants from seeds, each of the new plants will be genetically identical to the mother plant. "Hardwood" cuttings are taken during the winter dormant season. "Softwood" cuttings are taken from succulent new growth in late spring. "Semihardwood" cuttings are taken from partially matured new growth in the summer. Timing will depend on the type of plant being rooted. Cuttings should be 4 to 8 inches long, depending on the plant, and are best taken from the youngest growth of healthy, vigorous shoots. Strip the leaves off the bottom 60–75 percent of the cutting. Some species root easier if the cuttings are wounded by removing two thin slivers of tissue on opposite sides of the bottom of the cuttings. Some recommend dipping the cuttings in rooting hormone powder or willow water. Aspirin dissolved in water has been reported to work. Our tests have shown that most of these are not really worth the trouble on most plants. What does work, believe it or not, is saliva. Unless working with a poisonous plant, put the base of the cutting in your mouth. Stick the cuttings into pots or flats filled with some type of rooting medium: peat moss and perlite, sand, compost, etc. The best medium for propagating most plants is a compost-based potting soil.

Keep the cuttings warm and moist and in bright light until they form roots. It usually helps if you cover them with clear plastic such as dry cleaners' bags to keep the humidity up. It will help greatly if you have a greenhouse with a mist system to maintain 100 percent humidity. Cuttings will root in two to eight weeks, depending on the species, at which point you can dig them up carefully and pot them individually.

Softwood Cuttings

Softwood cuttings, also called slips or greenwood cuttings, are taken from vigorous growing plants at a stage when the stem breaks with a snap when it is bent. Young, healthy, vigorous plants provide the best cuttings.

Late spring and early summer are the best times to take cuttings. Plants are making their fastest growth at that time, and the potential for root growth is the best. No cutting should be allowed to completely dry out, but softwood cuttings are especially perishable. They are best taken in the cool morning and kept in water for a half hour or so before they are stuck into the medium.

In making a softwood cutting, cut it from the plant at an angle rather than straight across. Softwood cuttings 6 to 10 inches long root better than 3- or 4-inch-long cuttings. More reserve energy is stored in a larger branch, which aids initially in faster, heavier rooting and in better growth once rooting has occurred.

Herbaceous Cuttings

Houseplants and perennials, as well as some annuals, are propagated by herbaceous cuttings. The cutting should be made in the spring through summer. They are usually 2–4 inches long, cut diagonally at the end just a little below a node, and plunged in cold water for half an hour unless it is a milky-juiced cutting or a geranium, which should be allowed to become dry to the touch before being planted. Put under glass or plastic after placing in the propagation medium.

Leaf Cuttings

Many houseplants can be propagated by merely starting leaves, with or without the leafstalk. Plants such as the African violet, sedums, hens and chickens, artillery plant, and peperomia can be started by simply cutting off a leaf with its stem attached and burying the stem in a propagation medium.

Long, narrow leaves, such as those of the sansevieria, may be sliced into 1–2-inch pieces and stuck vertically in the medium. Keep the pieces top side up.

Hardwood Cuttings

Cuttings from deciduous woody plants are hard to root unless kept under mist spray. Hardwood cuttings are taken when the wood is dormant. Because these cuttings require very little equipment, they are convenient for home gardeners. Grapes, willows, spireas, cotoneasters, shrub roses, hydrangea, honeysuckle, mock orange, and numerous vines can be started from hardwood cuttings. Take cuttings anytime from early spring when buds are beginning to open till fall, which is the most commonly used time.

Malcolm Beck (the founder of Garden-Ville Company, San Antonio, Texas) accidentally discovered that grapevines, especially mustang (*Vitus rotundifolia*), rooted best if cuttings were taken in early spring when buds were breaking pink. He suggested this to a vineyard operator who tried this and got his best rooting rate ever. Howard experimented and discovered the same with grapes and figs, but other types of hardwood cuttings didn't work as well when taken at bud break. This is when the most energy is in the buds and tip growth of the plant.

For the fall cuttings, select only healthy wood that grew the previous summer. Cut the top end on a slant, slightly above a bud. Cut the bottom end on a slant also to expose more cut area to form roots. Make all the cuttings of each variety the same length, between 5 and 12 inches, and tie them in small bunches. Bury the bundles in slightly moistened sand or vermiculite and store where they will stay cool but won't freeze. By spring, the cuttings begin to form a thick callus on their bottom ends. Dip the callused ends in a rooting powder or soak in willow water and plant them in light, rich soil in a protected spot with morning sun and afternoon shade. As soon as the roots have started, give them some liquid natural organic fertilizer such as Garrett Juice.

Transplants

Flats or pots of new plants always need to have a good soaking before they are taken outdoors. We like to soak the plants in a bucket of water with seaweed until they are saturated before planting time. Using Garrett Juice in the water is also a good technique. Plant roots should be sopping wet and planted into a moist bed.

Pinching the lower leaves off lettuce, cabbage family plants, and tomato transplants is normally a good idea. Do not pinch the leaves off eggplant, peppers, or any vine crops. To protect young plants from cutworms, slugs, and snails, sprinkle a healthy amount of diatomaceous earth around the plants after planting. Crushed hot pepper also works.

Nursery-grown transplant showing all the wrong ways to grow plants—in peat moss with perlite or foam pellets, which leads to a dead, dry, and unhealthy plant showing no biological activity. Photograph by Howard Garrett.

Tree Planting

For the most part, trees are tough, durable, and easy to plant and transplant if treated in a sensible and natural way. It's a fact that trees are the easiest of all plants to plant and manage with organic techniques. Companies in the tree-growing business are severely penalizing themselves if they are not using organic techniques. Healthy, biologically active soil is the key.

To plant any tree (shade, fruit, big, little, native, or introduced) here's the plan:

1. **Dig a Wide, Ugly Hole.** Dig a very wide, rough-sided hole, three to

Tree planting—the right way. Photograph by Howard Garrett.

Tree planting—the wrong way. Photograph by Howard Garrett.

Bare-root planting techniques. Photograph by Howard Garrett.

four times wider than the tree ball, especially at the soil surface. Square-shaped holes also work. The point is to prevent the roots from circling in the hole. In other words, do not dig small, smooth-sided holes. The width of the bottom of the hole isn't important, but the depth of the hole should be exactly the same as the height of the ball. Measure—don't guess. It's better to dig a little shallow rather than too deep. If a little bit of the ball is sticking out of the ground after planting, that's okay, but when you overdig and have to put backfill under the ball, the tree can settle and drown. If you set the ball too low in the first place, that can be even worse. It's best to set plants 2–3 inches higher than ground grade (even higher with larger pots) to allow for settling and being planted too low in containers.

2. **Run a Perk Test.** When time allows, dig the hole and then, before planting, fill the hole with water. Plant the next day if the water has substantially drained from the hole. If it hasn't, you need to find a different place, amend the soil with additives, or improve the drainage of the area by adding drain lines or altering surface drainage.

3. **Treat the Root System.** Container-grown trees often have root-bound balls. If so, soak them in water for one to two hours. After the roots have become thoroughly hydrated, carefully unwind the roots, treat them with mycorrhizal fungi gel, spread the roots out to their proper radiating positions, and cover the roots with disturbed native soil. Balled and burlapped trees don't always need this treatment. Not only do tightly bound roots have great difficulty breaking away and growing into the surrounding soil, they also prevent moisture from getting into the rootball. In many cases, trees will live and appear to grow fairly well but will later snap off at the ground because of circling and girdling roots.

4. **Backfill with Existing Soil.** Backfill with the soil that came out of the hole. No bark, no peat moss, no compost, no foreign soil, and no fertilizer goes into the backfill. If you are planting in soils that have been highly disturbed, add some mycorrhizal inoculum to the backfill material. If the backfill is softer than the surrounding soil, a "pot effect" is created in the ground, which makes proper watering difficult and encourages roots to circle in the hole. Circling roots can eventually kill the tree. Settle the soil with water—don't tamp—no feet, two-by-fours, or anything else. Simply let the weight of water settle the soil naturally.

5. **Mulch the Top of the Ball.** After the backfill is settled and leveled with the surrounding grade, cover the disturbed area with a 1-inch layer of a 50/50 mix of compost and lava sand. Acceptable lava sand is red or black scoria that is the by-product of lava gravel and has a paramagnetism value of 800 cgs or greater. Zeolite at the same rate may be a better choice. Any volcanic rock material can be used if lava sand isn't available, and earthworm castings can

substitute for the compost. Finally, add a 3-inch or greater layer of shredded tree trimmings. Mulch should be 3 inches thick at the outer edge of the hole but decrease in thickness as it nears the trunk of the tree. There should be no mulch on the root flare or trunk of the tree.

6. **Do Not Wrap or Stake Trunk.** The exception is for heavy trees such as palms and evergreens. You may have to stake these for a week or two until soil is settled and new roots provide some anchoring against wind to keep tree from leaning. The ill-advised technique of staking trees was probably started years ago by those who planted bare-rooted trees and mistakenly put soft potting soil in the hole as backfill. Wrapping material around tree trunks probably got started because it looked important. Landscape contractors have admitted to me that brown paper wrap, tree stakes and guy wires, and even the troublesome watering rings are added to newly planted trees for no other reason than to impress the homeowner. Never mind the fact that all these additions are detrimental to the young trees.

Tree staking with wires, ropes, or cables cuts into the bark, or at least crushes the cambium layer (even if rubber hoses are used), and causes stress and long-term injury. Staking also prevents the natural movement of the tree in the wind, which prevents the development of trunk caliper and trunk strength.

I have asked many people, including contractors, landscape architects, and others, what the purpose is of wrapping gauze, paper, cardboard, or burlap around the trunks of newly planted trees. The answers range all over the place, but include protection from insects, diseases, lawnmower and string trimmer damage, and sunburn. Some tree wrappers admit that the only reason is that everyone does it. Look at the bark under some tree wrapping that's been in place awhile. You'll see that the cover actually encourages and protects insects and diseases and causes weak, shriveled bark—just like leaving a bandage on your finger too long. The only possible reason to cover tree trunks is the rare possibility of sunburn to the trunks of thin-barked trees that were grown in shade at the nursery. If you're worried about that, use a white wash of one-half white latex paint and one-half water. The tree will grow it off naturally. Tree Trunk Goop (see Appendix 1) could also be used. Trees planted properly don't need the stakes, the wrapping, or the expense.

7. **Do Not Build Water Dikes.** If you plant your trees correctly, these things aren't necessary. Supposedly these water-ring dikes form a dish that makes watering more efficient. The problem with that thought process is that when trees are backfilled with the existing native soil and a thick layer (3 to 5 inches) of mulch is tossed down on top of the disturbed area, moisture will stay in the root zone for a long time without the cost or inconvenience of the watering rings. If you build water rings around the trees, you have to tear

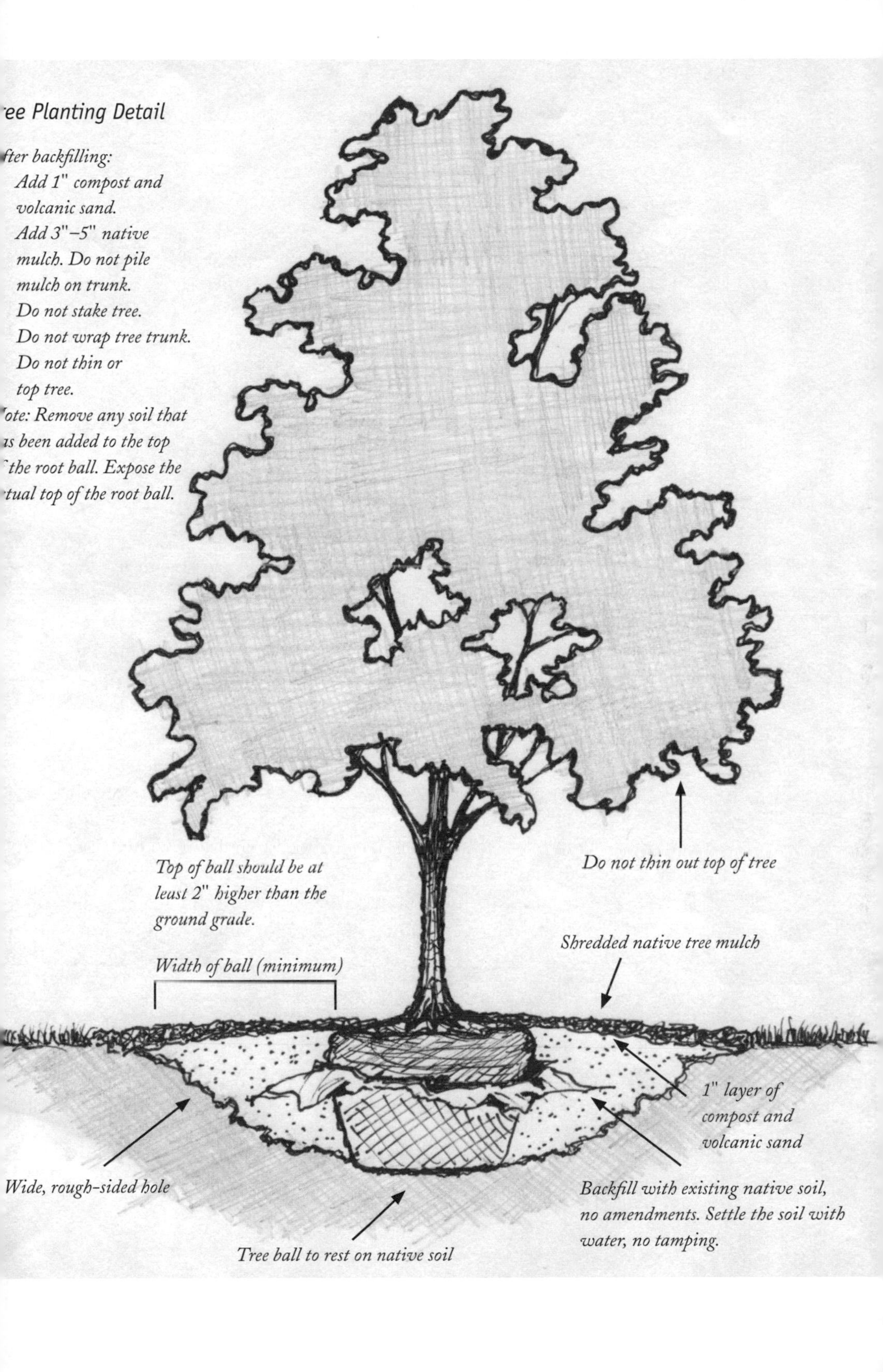
ree Planting Detail
fter backfilling:
Add 1" compost and volcanic sand.
Add 3"–5" native mulch. Do not pile mulch on trunk.
Do not stake tree.
Do not wrap tree trunk.
Do not thin or top tree.
ote: Remove any soil that
s been added to the top
the root ball. Expose the
tual top of the root ball.
Do not thin out top of tree
Top of ball should be at least 2" higher than the ground grade.
Shredded native tree mulch
Width of ball (minimum)
1" layer of compost and volcanic sand
Wide, rough-sided hole
Backfill with existing native soil, no amendments. Settle the soil with water, no tamping.
Tree ball to rest on native soil

them down at some point, or you'll have "watering bumps" around your trees forever.

8. **Do Not Cut Back the Top.** Thinning out the top of transplants and new trees is another old-time procedure that just doesn't make sense. Alleged experts still recommend cutting away as much as 50 percent of the top growth to compensate for root loss. I've planted lots of trees, including fruit trees, and they always establish and start to grow better when all of the limbs are left on the tree. Carl E. Whitcomb (the author of *Know It and Grow It*) has proven this with his plant research. Trees need foliage to collect sunlight, manufacture food, and grow. There are two exceptions in Texas—live oaks and yaupons transplanted from the wild. They *do* respond positively to a thinning of about 40 percent of the top. Why? We don't know. We haven't figured it out yet. It could be the reduction in weight, less wind resistance, or allowing more sunlight to the remaining leaves.

Some say that transplants have a large loss of fine and small feeder roots from being dug up, as well as loss of fungal colonization of the root hairs that were also left behind. This greatly reduces a plant's ability to absorb the nutrients and water required. Hence a reduction of top growth reduces demands on the root system. Most tree people disagree with this and have completely stopped thinning out and cutting back new or transplanted trees. Some arborists have been discussing the problems with both balled and burlapped trees and more so with container-grown trees and are leaning strongly toward recommending bare-rooting techniques again.

Mulching a Newly Planted Tree: Apply 2 to 3 inches of mulch around newly planted or transplanted trees and shrubs, and keep the mulch a few inches away from plant crowns or bark of plants and trees. Do not pile up mulch in deep cone-shaped mounds around tree trunks. This looks bad, holds moisture against trunks, and potentially leads to crown and stem rot. Extend mulch out 1 to 2 feet beyond the planting hole to allow for the season's root growth for trees.

Painting Tree Trunks: Painting or spraying the trunks of young fruit trees with white latex paint in the late fall may reduce the chance of freeze injury (southwest-side trunk damage). Paint from the ground up to just past the junction of the lower scaffold limbs. Painting tree trunks with a bright white latex paint to reflect the winter sun and minimize temperature fluctuations at the bark surface should be done sometime after harvest. Any inexpensive white latex paint will do. Nontoxic, low-VOC (volatile organic compounds) paints would be best.

Container Growing: We have interesting evidence that growing plants in solid compost or mulch works much better than we originally thought. Give the following materials a try: compost, shredded hardwood bark, and living mulch (composted native mulch). We have some very interesting anecdotal evidence from reader and listener reports that woody seedlings grow well in these coarse soilless media. There is also growing evidence that plants grown in containers have serious long-term problems because of circling roots in the containers.

Potting Soil

Tomatoes growing in soil not treated with mycorrhizal fungi (on left) and treated (on right). Photograph by Mike Amaranthus.

Potting soil—as opposed to native soil, loam, dirt, or landscaper's soil—is what should be used in pots no matter what the crop. It should be light, loose, well aerated, fertile, and full of biological activity.

We do not recommend peat moss–based potting soils. Peat moss is antimicrobial, meaning microbes don't grow well in it. That's just the opposite of what we want. Peat moss is good for storing bulbs or shipping food or other live material. Potting soil should not be sterile, as some recommend, but alive and dynamic. It should be full of microorganisms and have the ability to stimulate quick and sustained growth.

Indoor plants and outdoor potted plants should be planted in a well-drained organic potting soil. Excellent basic ingredients are compost, coconut fiber, and expanded shale. One excellent formula is as follows:

- 40 percent compost
- 40 percent coconut fiber
- 20 percent mix of lava sand, granite, zeolite, expanded shale, greensand, alfalfa meal, earthworm castings, and molasses

CHAPTER 4

Fertilizing

The natural way to fertilize is to feed the soil and let the healthy soil feed the plants. The key to natural fertility is biological activity in the soil. By-products of microbe feeding, microbe waste, and dead bodies of microbes are the true natural fertilizers for the soil and the plant roots. Helping this happen in an efficient and cost-effective way is the key to the commercial application of the organic program.

How Organic Fertilizers Work

Organic fertilizers do not have very high amounts of nitrogen, phosphorus, and potassium, as they do not need them. Very little of the toxic synthetic fertilizers actually makes it to the plant, because they are water soluble and most of the nutrients leach away to pollute waterways, hence high levels are required just to get a little bit into plants.

When fertilizing or adding mineral nutrients, it's important to think about balance. Healthy soils and plants have a balance of elements and ingredients. A proper fertilization program will help keep that balance intact. That's why it's important to avoid an overkill of the well-known elements nitrogen, phosphorus, and potassium. Synthetic fertilizers are not balanced. They have too much nitrogen, no carbon, and a poor collection of trace minerals.

Here's a good example. The following are the percentages of various elements in whole plants:

Oxygen	45 percent
Carbon	44 percent
Hydrogen	6 percent
Nitrogen	2 percent

Potassium	1.1 percent
Phosphorus	0.4 percent
Sulfur	0.5 percent
Calcium	0.6 percent
Magnesium	0.3 percent

Note the relatively low percentages of nitrogen, phosphorus, and potassium and the high percentages of oxygen, carbon, and hydrogen.

When buying fertilizer, remember how relatively unimportant nitrogen, phosphorus, and potassium are. Think in terms of providing the soil with those ingredients that will help maintain the natural balance. If the soil is in a healthy, balanced condition (which includes plenty of organic matter and air), nitrogen, potassium, and phosphorus will be produced naturally by the feeding of microorganisms, and relatively little will need to be added. Much of the N-P-K in synthetic fertilizer is not used by plants, but rather is wasted to volatilize and leach away to contaminate the environment.

The natural-organic versus inorganic-chemical argument continues. Organic proponents say only organic products should be used. The synthetic proponents have their argument of the value of higher N-P-K analysis and quicker availability. They base their recommendations on the theory that Justus von Liebig delivered to the British Association for the Advancement of Science in which he said that plants find new nutritive materials only in inorganic substances. Later, in 1843, the father of the N-P-K theory completely recanted and apologized for the error. Unfortunately, his gross mistake still lives today and is fraudulently used to justify the synthetic fertilizer industry.

Synthetic "weed and feed" fertilizers should be avoided. Photograph by Howard Garrett.

Most of the fertilizer elements called chemicals occur naturally in nature. In fact, that is where humans discovered them. Among these are ammonia, ammonium, ammonium sulfate, nitrites, nitrates, potassium sulfate, calcium phosphate, and urea. Seldom are these chemicals found in a pure state, however. They are almost always bound up in rock or in an organic form with other elements. They may also be found in a state of transition.

Human-made synthetic fertilizers always have a high total N-P-K, from 20 to 60 percent or more. The total N-P-K of natural organic fertilizer blends will always be low, usually no higher than 14 percent.

The balance of the ingredients in the synthetic fertilizer bag, aside from the total N-P-K, is usually inert filler or, possibly, chemicals that aren't needed. If the label says 10-10-10 (30%), the other 70 percent is filler. Some companies use hazardous waste or materials from Superfund sites as fillers. The state of Washington has measured the contaminants in many brands and posted this information on their Web site. The balance of the ingredients in the natural organic fertilizer bag beyond the total N-P-K is all necessary nutrients and organic matter. Organic ingredients are materials that came from a once-living entity—plant, animal, or a blend of both—which tells us that every ingredient is important and is in correct proportions to feed and sustain the next generation of life. Other ingredients in natural fertilizers are the rock minerals, sugars, and living organisms.

Many synthetic fertilizers are labeled "complete." This is really a false statement. It takes much more than a few chemicals to maintain a healthy soil and grow healthy plants. For example, there is very little, if any, carbon in a bag of synthetic fertilizer. When a plant or animal body is analyzed, one of the most abundant elements in it is carbon, in the form of carbohydrates that can be used to produce energy.

For a plant to be properly fed, whether with synthetic or natural fertilizer, the microbial life in the soil must first process it and release it in the correct amounts that are perfect for the plant to absorb. But for the microbes to perform this service, they must have energy. They are not in the presence of sunlight, nor do they have chlorophyll like higher plants, so the microbes must get their energy from decaying plant or animal matter (carbon-containing material) stored in the soil. Another problem with synthetic fertilizers is that a large percentage of the products volatilizes and leaches away to pollute the environment. They are a significant source of air and water pollution. The soluble fertilizers also prevent certain microbe spores from germinating.

A bag of natural organic fertilizer has all the carbon energy to meet the needs of the soil microbes. A bag of synthetic fertilizer usually has no energy. If organic matter is not already present in the soil, the chemicals can quickly become stressful, even toxic, to the plants. This causes plants to be susceptible to disease and insect problems. Synthetic fertilizers are harsh chemical salts; we add salts to pickles, canned vegetables, salted meat, etc., to *kill* microbes. The same thing happens in the soil. We do, however, recommend sea salt as an excellent soil amendment.

Natural organic fertilizers are believed to be slower acting than the synthetics. This is true to a degree, but it's not a bad thing. Having a lower N-P-K analysis, slower-acting natural organic fertilizers can be used in higher volumes around plants without danger of burning them. However, there are some natural organic fertilizers that are fast acting. Bat guano and fish meal can show results as quickly as the synthetic fertilizers, but they are still slower to burn than synthetics and last much longer in the soil. Liquid fish emulsion and seaweed products are also fast acting.

Unless synthetic fertilizers are impregnated or coated with a microbe inhibitor and/or some substance to keep them from quickly dissolving, they must be used very cautiously, especially in sandy soils, because they can burn the roots, then quickly leach beyond the reach of the roots. They generally end up polluting the soil and water supply because they are too quickly dissolved and moved out of the soil. This is less a problem in heavy clay soils or any soil with a high organic and humus content.

Synthetic fertilizers that are blended to perfectly fit a given soil and then used in the correct season and correct amounts can do nothing more than grow a plant. They do not build or sustain a healthy soil and often have negative effects on the beneficial microbes, earthworms, other soil life, and plants themselves in addition to polluting the environment.

Natural organic fertilizers contain the energy (carbon) and trace minerals that continually build the soil's fertility, crumb structure, and water-holding capacity and provide food for all the beneficial soil life. They condition the soil and contribute to the hundreds of other as-yet-unknown things that cause a plant to grow healthy and perfect. Only healthy and perfectly grown plants can properly feed and support a healthy and perfect animal or human life.

High-nitrogen fertilizers have other problems. They can cause severe thatch buildup in lawns by forcing unnatural flushes of green growth. That's why mechanical thatch removal programs are often recommended for synthetically maintained lawns. Organic lawn-care programs take care of thatch problems naturally, as the living microorganisms feed on the grass clippings and other dead organic matter to prevent a thatch buildup.

High-nitrogen fertilizers such as 15-5-10 (or even higher) are still being recommended by many in the farming, ranching, and landscaping businesses. These amounts of nitrogen, phosphorus, and potassium are unnecessary and even damaging to soil health.

Synthetic fertilizers are fast-acting short-term plant growth enhancers and are responsible for several problems: (1) deterioration of soil structure, creating hardpan soil; (2) destruction of beneficial soil life, including earthworms; (3) changing nutrient and vitamin content of certain crops; (4) making certain crops more susceptible to diseases and insect pests; (5) preventing plants from absorbing needed minerals. In other words, using synthetic fertilizers is the best way to obtain short-term results and long-term damage to the soil, the groundwater, and the health of our animals and ourselves.

Synthetic N-P-K fertilizer is just that, a synthetic form of nitrogen, phosphorus, and potassium with possibly some form of trace or minor elements blended in. Most are highly soluble, so they can leach past the roots into the water table to pollute well water or run off into streams.

Chemical N-P-K does not contain any energy, so if the microbes are to process it into a better form for plant use, they will have to use energy from the soil, which usually depletes the soil energy reserves (destroys humus) faster than nature can replace it. Once the soil energy is depleted, the pure N-P-K doesn't get processed into the microbe bodies and exudates, which means the plants will pick it up in wrong proportions, causing them to grow weak and unbalanced. Then nature's scavengers, the pest insects and diseases, move in to kill the weak plants and return them to the soil for energy food for the microbes. Several studies have shown that pest insects are actually attracted to plants fed synthetic fertilizers. Another problem is that many synthetic fertilizers acidify the soil to the point that lime must be added, which then cements the soil into hardpan. Additionally, nitrogen in the form of nitrates favors weed growth because perennials, shrubs, and trees do not want their nitrogen in a nitrate form.

The N-P-K in natural organic fertilizers comes impregnated in carbon compounds that represent the energy the microbes need to process it properly. The microbe processing happens when moisture and temperature are perfect for plants. Also, no soil energy is robbed from other soil life, and it can't leach away. Organic N-P-K comes from a once-living entity, so it contains not only N-P-K but also all of the minor trace elements in correct proportions to sustain the next generation of life.

We have had a considerable amount of feedback as to how well molasses works as a fertilizer. Similar reports have come in about the advantage of applying white sugar. The carbon is the key; these sugars are providing carbon for the feeding of microbes. One gardener has reported that his landscaping, gardens, and turf look as good under a sugar-water spraying program as they did under a full fertilizer regime. The fertility value provided by the waste and dead bodies of microbes is the true natural plant food.

No matter what program is being used, there is a foolproof way to determine if the nutrients are getting into the plants. The available nutrients and trace minerals are directly related to the sugars or complex carbohydrates in the plants. A simple device called a refractometer tests for these sugars (see page 63).

Foliar Feeding

Foliar feeding is a method of fertilizing through the foliage of plants with a liquid spray. A mixture of liquid seaweed and fish emulsion is the most common recommendation given by organic practitioners, but there is actually a better recommendation now. It is Garrett Juice, which is a mixture of compost tea or liquid humate, natural vinegar, molasses, and seaweed.

Foliar feeding has been used since 1944, when it was discovered that plant nutrients could be leached from leaves by rain. Experiments soon proved that nutrients could also enter the plant through the foliage. It is still somewhat of a mystery as to exactly how the nutrients enter the plants through the foliage—through the stomata or right through the cuticle of the leaves?—but it is agreed that it works and it works quickly and efficiently. There is also evidence that nutrients can be absorbed through the bark of trees.

T. L. Senn of Clemson University wrote about foliar feeding in detail in his book *Seaweed and Plant Growth*, in which he explains the power of seaweed as a fertilizer and root stimulator and how foliar feeding can be used to supplement a fertilization program. Foliar feeding is several hundred times more efficient than soil fertilization, according to Senn, and organic foliar sprays are the most effective, since the nutrients are in a more balanced proportion for plant growth.

Foliar sprays of organic products activate plant growth and flower and fruit production by increasing the photosynthesis in the foliage, the translocation of fluids and energy within the plant, and the uptake of soil nutrients through the root hairs as well as by stimulating microorganisms in the soil and on the leaves. Foliar feeding increases the efficiency of all the natural systems in the soil and plants. The end result is a bigger, stronger, healthier plant with increased drought and pest resistance.

Here are some points to remember when using foliar sprays:

- Less is usually better in foliar sprays because light, regularly applied sprays are better than heavy, infrequent blasts on plant foliage.
- Mists of liquids are better than big drops.
- Young, tender foliage absorbs nutrients better than mature, hard foliage, so it is best to foliar-feed during the periods of new growth on plants.
- Sugar and molasses added to spray solutions can stimulate the growth of beneficial microorganisms on leaf surfaces and in the soil.
- The stimulation of friendly microbes helps to fight off harmful pathogens.

- Well-timed foliar feeding on food crops will increase their storage life.
- Spraying on damp, humid mornings or cool, moist evenings increases the effectiveness of foliar sprays. The least effective time to foliar-feed is during the heat of midday. Small openings in leaves called stomata close up during the heat of the day or as a result of other stress so moisture within the plant is preserved. The very best time to get the indirect pest control may be dusk, so that the liquid stays on the plant leaves as long as possible during the night. Some would argue that this increases diseases, but we are not sure we agree with that. Gases and liquids are best absorbed through the leaves during this period. The fluids that are moving down the plant and roots are at their peak during the cooler hours of the evenings.

Product Information

Our best recommendation for a foliar-feeding material is aerated compost tea or Garrett Juice in the formula containing aerated compost tea. Formulas for ready-to-use and concentrate mixtures can be found in Appendix 1.

Micronized Products

Micronized products are made by mechanically pulverizing them into an extremely fine texture. Properly micronized finished products have the consistency of talcum powder. The particles range in size from 5 microns to 75 microns. Their advantage is the tremendously enlarged surface area of the product, which enables plants to access the nutrients more efficiently. These products can be mixed dry into potting soils and beds or mixed with water and sprayed or drenched into the soil. This is an excellent choice when the use of compost tea would cause logistical problems. As the particle size gets smaller, the total surface area increases greatly, thus allowing more microbes to attack and dissolve the particle.

Microbe Products

Products are now available that contain many species of beneficial soil bacteria. Some will pull nitrogen out of the air in soil and make it available to feed plants. When sufficient nitrogen-fixing bacteria are in a soil, the need for fertilizer is reduced. Other bacteria decompose organic matter and break down pesticides and other toxic residues in the soil. Soil bacteria reduce soil compaction by improving soil structure, creating microscopic spaces in the soil to hold air and water. Some soil bacteria attack and kill soil pathogens. Other products contain beneficial fungi that grow on the roots of plants, increasing the plants' ability to pull nutrients and water from the soil. Root growth is greatly enhanced when transplants and seeds are exposed to the products prior to planting. These living products can also be applied to growing plants, but they have to get into the soil and have contact with the roots to work. Mycorrhizal fungi production works better in soils rich in organic matter with good structure and aeration. The best microbe products use a blend of bacteria and fungi. Examples of companies with these products include Alpha BioSystems and Bio-S.I. (see Appendix 2).

Refractometer

A refractometer is a terrific instrument that every horticultural professional should have. Although it has been used in the grape-growing industry for many years, it needs to be talked about more often. It is an optical instrument used to measure the amount of dissolved solids in a liquid. It shows the sugar levels in plant tissue.

A refractometer gives growers the Brix, which is the relative "sugar weight" or percentage of a sample compared to distilled water. It can be used on all

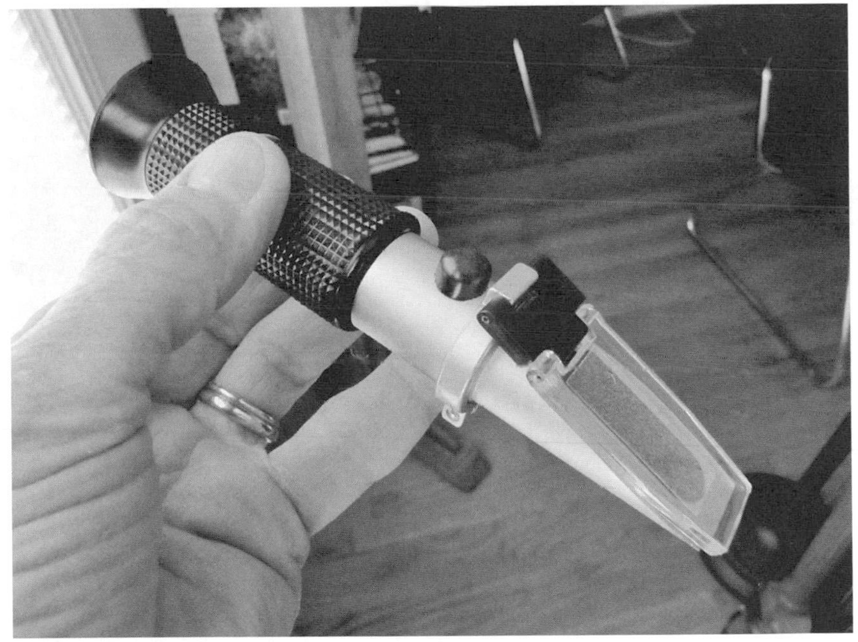

Refractometer used to measure (in degrees Brix) the sugar in plant tissue. An excellent tool to learn what nutrients are getting from the soil into plants. Photograph by Howard Garrett.

plants and all fruit. The refractometer works much like a prism—it reacts differently to light (by giving a reading on a scale), depending upon the amount of sugar that is available in the liquid sample held between the daylight plate and the main prism assembly. As light moves through different media, its speed changes. When light comes into the refractometer prism at an angle, it bends as it moves from one material to another. It bends toward the denser material. Due to optical density, light slows down when passing through a dense liquid. Simply place a sample on the glass and close the daylight plate (with no bubbles or air spaces showing). Light enters from all angles.

Refractometers are traditionally used to test ripening fruits and vegetables; however, they can also be used for plant sap analysis. Sap is squeezed from fresh plant tissue (leaves or petioles) and tested for Brix or sugar content. The higher the sugar, the healthier the plant and the better the taste. Food products with higher Brix readings are healthier for people and animals, and plants with higher Brix readings have few to zero insect and disease problems. Devices are available from places like Pike Agri-Lab Supplies, Inc. (http://www.pikeagri.com).

CHAPTER 5

Pest and Disease Control

The primary cause of insect pest and disease problems is the regular use of high-nitrogen synthetic fertilizers and toxic chemical pesticides. These products treat symptoms rather than the true problems and injure more beneficial life than the targeted pests. It's as simple as that! Research has shown that pests are actually attracted to weak, unhealthy plants that have resulted from using toxic chemicals and to plants fertilized with artificial fertilizers.

The Real Purpose of Toxic Chemical Pesticides

Toxic chemical pesticides (e.g., synthetic neurotoxins and other killers) have many problems, but they do have a purpose: to attempt to keep unhealthy plants alive. When troublesome insects attack and start to eat the foliage or stems or roots of ornamentals and food crops, synthetic pesticides can be sprayed to knock down the pests. Besides helping to keep unhealthy plants around, they also kill beneficial insects and microbes in addition to the plant-eating pests. They are toxic to pets and humans. They contaminate the air, soil, and water. They are costly and time-consuming to apply. But, they do make money for the people who manufacture and sell them.

Insect pests on plants are related to plant stress. Unless the cause of plant stress is solved, pests tend to repopulate and often in greater concentrations than the original infestation. Now more pesticides have to be sprayed to control new infestations, which starts a vicious cycle. That doesn't make anyone happy except those who sell chemical products. But there's more. It's common for those gardeners who use toxic chemical pesticides to also use synthetic fertilizers. These unbalanced fertilizers are one of the primary causes of plant stress. The greater the stress, the greater the pest populations, and the greater need for more pesticides. Increased pesticide spraying increases plant stress.

Plant diseases are usually caused by four major types of living microorganisms: fungi, bacteria, viruses, and pathogenic nematodes. Diseases are basically an imbalance of microorganisms. Plant diseases are sometimes hard to identify, since the results of infection are more visible than the organisms themselves.

All organic products help control disease. When soil is healthy, there is a never-ending microscopic war being waged between the good and the bad microorganisms, and the good guys usually win. Disease problems are simply situations where the microorganisms have gotten out of balance, but if allowed to do so, the good guys will control the bad guys. When pathogens are brought into their proper proportions, they are no longer troublesome. In most cases, they become beneficial at that point. When the microbes are in the proper natural balance, even the so-called pathogens are beneficial. Natural disease control products don't kill disease organisms, but rather stimulate beneficial organisms, such as Trichoderma, a bacterium that eats pathogenic fungi. Diseases like brown patch and other soil-borne diseases are easily controlled by healthy, biologically active compost or just plain cornmeal. These organic products don't hurt the other beneficial organisms at all. The toxic chemical fungicides, on the other hand, kill indiscriminately—they get the good guys *and* the bad guys, creating a semivacuum. Nature doesn't allow that, so guess what? The pathogens grow back faster than the beneficial microbes. Plants weakened by disease are sitting ducks for other pest attacks.

Insects and diseases are nature's cleanup crews. Their job is to move in and take out unfit plants. Preventing that natural process without solving the cause of the stressed plants serves no one but those who sell the toxic pesticides. The purpose of the synthetic pesticides? To keep sick, unhealthy plants alive so more toxic rescue chemicals can be sold!

Pests and Their Organic Remedies

Insect Pests in General—improve soil and plant health, water properly, treat with beneficial insects and microbes first, treat with food products that repel instead of kill second (garlic, pepper, cinnamon, etc.), and as a last resort use the killing products (natural diatomaceous earth, orange oil, neem, etc.).

Diseases in General—improve soil health, water properly, and spray liberally with compost tea and similar products. Infestations can be stopped with cornmeal and cornmeal juice spray (see Appendix 1), potassium bicarbonate, neem, hydrogen peroxide, and Bio Wash (to clean plants).

Pest Control Products to Avoid in General—all toxic synthetic products and all toxic organic products.

Acceptable Organic/Natural Pest Control Products

The list of acceptable products is continually changing as companies that provide them go in and out of business. For the most current listing of acceptable pest control products for an organic program, please refer to this Web site: http://www.dirtdoctor.com/view_org_research.php?id=112.

An up-to-date listing of unacceptable pest control products can be found at: http://www.dirtdoctor.com/organic/garden/view_org_research/id/68/.

Insect Pests

Ants: There are many different ants, including carpenter ants, fire ants, sugar ants, leaf-cutter ants, and pharaoh ants. Solutions for ants indoors include natural diatomaceous earth, boric acid, cinnamon, and baking soda. Fire ant mounds should be treated with the Fire Ant Mound Drench solution (see Appendix 1 for formula). Treat the site with beneficial nematodes and go organic. The competition of microorganisms, insects, lizards, frogs, toads, birds, and even plants is not enjoyed by fire ants. Control carpenter and other house ants with sweet baits that contain small amounts of boric acid. Orange oil or D-limonene sprays will also kill them. Abamectin baits can be used for ants that continue to invade.

Ants, Acrobat: Like all ants, this species has a complex life cycle, developing from eggs into white legless larvae and then pupae before emerging as adults. Development from egg to pupa takes place within the nest, and immatures are rarely seen. Like all ant species, acrobat ants produce winged individuals known as swarmers, which are fertile adult males and females whose only function is to reproduce and found new colonies. They do not forage for food, bite, or sting. The males (drones) and females (queens) emerge, take flight, and mate while in flight. Females land, shed their wings, and seek soft soil in which to create a nest. The males die shortly after mating. Acrobat ant swarmers usually emerge in the fall, although flights have been observed as early as June. Acrobat ants can be nuisance pests, and workers can be aggressive if disturbed and may sting or bite. Some species also produce a foul odor. This ant prefers to forage outside. Workers may travel over 100 feet from the nest in search of food. Acrobat ants feed on a variety of foods, including sweets and other insects. They have frequently been observed feeding on termites. Structural damage associated with this ant is minimal. Occasionally,

Trichogramma wasp—one of the most efficient and cost-effective beneficial insects that can be purchased (on cards, as here, or strips) and used to control many species of lepidopterous insect pest larvae. Photo shows moth eggs containing the entire trichogramma wasp life cycle. Photograph by Howard Garrett.

Hover fly—an important insect pollinator whose larvae are effective predators helping with pest control. It is often misidentified as a wasp. Photograph by Howard Garrett.

this ant will expel fine frass (sawdustlike excrement) from the nest, which causes homeowners concern. Acrobat ants do not attack sound wood. The presence of these ants in structures is often indicative of a moisture problem related to a leak or condensation.

Ants, Carpenter: Wood-eating ants can be controlled with boric acid wood treatments or dusting indoors. Wall cavities can be treated with a mix of 90 percent natural diatomaceous earth and 10 percent boric acid. Do not use boric acid outdoors. It can cause severe soil toxicity. Orange-oil products are effective at killing and continuing to repel the pests.

Ants, Crazy Rasberry: These ants are a relatively new pest that has become a problem in the South. They do significant damage to electrical devices. They can be controlled with essential oil products or easily with natural diatomaceous earth.

Ants, Fire: Toxic chemical products like Dursban, diazinon, orthene, Amdro, and others do not work. The fire ant problem has been caused by using toxic chemical pesticides. If we had allowed them to be controlled by the native fire ants and other insects rather than broadcasting chemicals from airplanes, we probably wouldn't have the problem we have today.

Use the organic three-step approach. (1) Drench mounds with an orange-oil-based mound drench product (molasses, compost tea, and orange oil; see Appendix 1). Natural diatomaceous earth can also be stirred dry into mounds. (2) Apply beneficial nematodes to the site. (3) Use a total organic program. The competition from beneficial microbes and insects, including other ant species, is very important. Controlling fire ants with chemicals is impossible. That has been proven over the past forty years.

The first step is to treat with the drench the fire ant mounds that are the most visible and the biggest problem. The material has to be poured into the center of the mound so that it goes to the bottom quickly and kills the queens before it is splashed on the rest of the mound. This drench will kill the ants 100 percent if it coats their bodies. Fire ants do not like heavy biodiverse planting and/or microbially active soil. For this reason, broadcasting dry molasses will often end fire ant problems.

Ants, Harvester (Red Ants)*:* These are very interesting and beneficial ants, but be careful: they have a powerful bite and a powerful sting, though they bite only when you disturb their mound. They are not as aggressive as fire ants, but their sting is more painful than a fire ant sting. When treating for fire ants, avoid harming these ants. They should be protected. To move a mound of these beneficial ants, cover it with shredded mulch. The ants will relocate to a spot where the opening to the mound is again in full sun.

Ants, Sugar (Pharaoh Ants)*:* Small indoor pest ants that can be controlled with natural diatomaceous earth, orange-oil spray, boric acid, and sugar baits. Baits containing apple mint jelly and a small amount of boric acid are effective. Powdered cinnamon is an excellent repellent.

Ants, Texas Leaf-cutting: Aggressive "cut ants" strip the foliage of plants and take it into large underground nests where the organic matter is composted to grow a fungus that is used as their primary food source. Control is difficult. Biodiversity helps, as does keeping all bare soil mulched. Drenching the mound area after building a dam around it is a lot of trouble but seems to work. Add orange oil for better results. Drench applications of wettable sulfur work well because the sulfur fouls the fungus being grown by the ants. It's best to do this work during cool weather. Repeated applications of the Fire Ant Mound Drench formula or natural diatomaceous earth have been reported to work. Essential oil products have also been reported to be successful.

Ants, Odorous: Like all ants, odorous house ants live in social colonies. These colonies are made up of workers and reproductives. Swarmers are sometimes misidentified as termites. Ants can be identified by the front wings being larger than the hind wings. Wings on termites, however, are considerably longer than the body and both wings are the same size. Odorous house ants are very opportunistic and can nest in many different places, both indoors and out. Odorous house ants forage both night and day and eat many types of foods. They eat live and dead insects but are also very attracted to sweet foods. They especially like the honeydew that is produced by aphids and mealybugs. Many colonies of odorous house ants tend or herd aphids and mealybugs to collect the honeydew they excrete.

Aphids: These sucking insects can destroy the tender growth of plants, causing stunted and curled leaf growth and leaving a honeydew deposit. The presence of these insects indicates plant stress. Spray foliage with a strong water blast, garlic-pepper tea, or Bio Wash to clean plants. Add Garrett Juice for even better control. Release beneficial insects—ladybugs, brachonid wasps, green lacewings. Giant bark aphids require no treatment at all. Apply the Sick Tree Treatment (see Appendix 1). Trap crops like Mexican milkweed or other *Asclepias* are also helpful.

Armyworms; Cabbageworms: Spray Bt (*Bacillus thuringiensis*) products with 1 ounce of molasses added per gallon of water. Spray at dusk. Release trichogramma wasps in the early spring and encourage and protect native wasps. Dusting with natural diatomaceous earth will also help. Native wasps eat many common caterpillar pests.

Note: The biological pesticide *Bacillus thuringiensis* is a bacteria applied as a spray to kill caterpillars and sold under a variety of names such as Thuricide, Dipel, Bio-Worm, and others. I rarely recommend Bt anymore and never as the first choice. I would personally never use it because other things work as well or better such as release of trichogramma wasps as a preventative and EcoSMART or EcoEXEMPT as curatives. In addition, Bt kills butterflies and beneficial moths. The good side of all this is that more and more exposure is being given to the problems related to GMOs and products that contain the Bt gene.

Bagworms: Bagworms are a common pest of ornamental trees and some shrubs. They will prey on many different species of plants such as cedar, juniper, cypress, etc. In the larval stage, they can defoliate trees. They can be controlled with *Bacillus thuringiensis* (Bt) (see warning above) in the spring. Handpicking the bags is the only control once the bags have been attached to the plants. Trichogramma wasps can also help to control problem infestations. Apply the Sick Tree Treatment. Release trichogramma wasps and green lacewings in the early spring. As a last resort, spray a Bt (*Bacillus thuringiensis*) product and add 2 tablespoons (1 ounce) of molasses per gallon of spray.

Ball Moss: Not a serious problem for trees other than cosmetic. It can be controlled by spraying baking soda or potassium bicarbonate at ⅓ to ½ cup per gallon of spray. Like mistletoe, it does tend to attack trees as their health declines and the canopy thins. Apply compost mulch and the entire Sick Tree Treatment for long-term control.

Bees: First of all, we need to strongly remind you that all bees are beneficial. But if they are swarming and a problem, they can be repelled with various sprays, including citronella, citrus oil, and garlic-pepper tea. To treat hives, wait until dark. Try to run them off without hurting them and then caulk the cracks and crevices. For the proper environmental control, contact the beekeeper club or society in your area. They will usually come and relocate them or give you advice on control. Problem bees can be killed with soapy water. Toxic poisons should never be used, especially to kill colonies. Other insects forage the poisoned honey late in the season and will kill their own colonies.

Our pollinator populations are seriously on the decline because of the use of toxic chemical pesticides.

Beetles: Some beetles attack plants that are in stress. Many adult beetles eat plant foliage and can destroy plants completely. An effective solution for destructive beetles is dry natural diatomaceous earth followed by a spray of Garrett Juice plus orange oil at 2 ounces per gallon of spray. Garlic tea is an even less toxic control. It's important to remember that many beetles are beneficial and only eat problem insects. Apply beneficial nematodes to the soil for additional help.

Borers: Borers are the larvae of beetles, moths, and some flies that attack sick softwood trees and various other trees that are in stress. Some of the causes of tree stress include pruning wounds, high-nitrogen synthetic fertilizers, physical damage, sun scald, soil compaction, chemical damage, and planting ill-adapted trees. Adult beetles will eat tender terminal growth and then deposit their eggs in the base of the tree. Eggs hatch into larvae and bore into trees and tunnel through the wood until the tree is weakened. Orange oil mixed 50/50 with water and painted on trunks or injected into holes will kill them. Beneficial nematodes applied directly in holes will usually kill active larvae, but keeping trees healthy and out of stress is the best prevention. A generous amount of diatomaceous earth at the base of susceptible trees will also help. Improve the soil health and the immune system of plants by using the Basic Organic Program. Applying beneficial nematodes to the soil and Tree Trunk Goop (see Appendix 1) to trunks is also effective.

Cabbage Loopers: Cabbage loopers are caterpillars and are the larvae of moths that are brown with silver spots in the middle of each wing. They can be killed with *Bacillus thuringiensis* (Bt) spray when the insects are young (see warning above). Add an ounce of molasses per gallon of spray. Spray late in the day, since these guys feed at night. Trichogramma wasps will also help control these critters. It's interesting that even the chemical "pushers" admit that the chemical insecticides are ineffective at controlling loopers.

Cabbageworms: Small green worms that eat leafy vegetable crops and other plants. Release trichogramma wasps early in the season. Spray with citrus-based products or a Bt product when worms are present. Spinosad is also effective but will kill honeybees.

Cankerworms: Larvae or caterpillars that hang on silk threads from trees. They do a lot of damage to early-season foliage. Wasps will usually control them. If not, use *Bacillus thuringiensis* (Bt) for heavy infestations (see warning above). Release trichogramma wasps early in the season and spray Spinosad as needed.

Casebearers: Small insects that damage young pecan nutlets. Trichogramma wasps are the best control. Release eggs at least every two weeks starting with leaf emergence. Some casebearer damage is helpful by reducing the over-abundance of pecans.

Caterpillars: The larvae of moths and butterflies eat plant foliage. Some are more troublesome than others. Release beneficial insects and especially encourage and protect native wasps. Release trichogramma wasps in the spring. Green lacewings are also quite effective. Citrus-based sprays can be used for direct kill. It is not appropriate to try to kill all caterpillars.

Chiggers: Microscopic pests related to ticks and spiders. The itching usually starts the day after you are bitten and lasts two to four days. Dust natural diatomaceous earth on infested area at 1 cup per 1,000 square feet. Broadcast sulfur at 5 lbs. per 1,000 square feet and increase the humus and moisture of the soil. Citrus-based products and light applications of sulfur control the critters. Vinegar or orange oil rubbed on bites will eliminate the itching. Nematodes will help break the breeding cycle.

Chinch Bugs: Chinch bugs are tiny black-and-white pinhead-size or smaller bugs. During hot, dry weather, chinch bugs can destroy unhealthy lawns. The lawns will look yellow, turn brown, and then die. A dusting of natural diatomaceous earth works in the hot, dry weather, as does Garrett Juice plus orange oil. Dusting sulfur also helps. This insect hardly ever attacks healthy, well-maintained grass and primarily attacks St. Augustine grass. Composted turf rarely if ever has chinch bug infestations.

Cockroaches: There are numerous cockroaches, but only a few really pose a problem. Cockroaches usually live outdoors and are nocturnal by nature. Roaches will enter a home or building through any crack or crevice. Solutions include natural diatomaceous earth; eliminating drips, leaks, standing water, and food sources; and sealing all openings. A light dusting of boric acid or placement of boric balls or natural diatomaceous earth indoors gives effective control. Spray with orange-oil products. Boric balls are baits made by mixing 10 percent boric acid, 70 percent flour, and 20 percent sugar. Add water and roll into balls or cakes. Abamectin baits are best for problem infestations.

Crickets: Crickets live in and out of doors; destroy fabrics such as wool, cotton, synthetics, and silk; and also attack plants. Their irritating sound is the primary objection, although they will eat tender sprouts of wildflowers and vegetables. Solutions include natural diatomaceous earth for outdoors and boric acid for indoor use. *Nosema locustae* is a biological bait for overall control. Orange-oil sprays will also kill them effectively.

Colorado Potato Beetle: An insect pest of potato crops. Treat with Bt-sd (*Bacillus thuringiensis* var. *san diego*) products at dusk per label instructions (see warning above). Garlic-pepper tea and diatomaceous earth also help. Encourage ground beetles and apply beneficial nematodes. Potato beetles overwinter as unmated adults. Handpicking the first to appear will soon eliminate them. No need to pick them all—just enough to upset the balance in your favor. Their natural enemies will soon wipe them out.

Coyotes: Use donkeys and llamas to protect livestock. To keep coyotes from eating watermelons, tape a strip of tinfoil to the watermelon with duct tape. The size of the foil can be as small as 4 inches square.

Crawfish: One of the most difficult pests to control. Crawfish are attracted to wet soil, so the only control that won't contaminate the soil is to eliminate the excess moisture by improving the drainage and reducing or eliminating irrigation.

Cutworms: These worms are the larvae of moths that damage young sprouts and older plants. Pour a ring of natural diatomaceous earth around each plant; bonemeal or rock phosphate will also help. Spray Bt (*Bacillus thuringiensis*) on young plants at dusk (see warning above). Add 1 ounce molasses per gallon. Collars are sometimes used around vegetable stems, but we think this is too much trouble. Cedar mulch and heavy applications of lava sand or gravel will also help. Fire ants do have a positive side—they will wipe out cutworms.

Elm Leaf Beetles: Wherever there is an American, Siberian, or cedar elm tree, elm leaf beetles can be found. They will eat and damage foliage and then move to the next tree. Trees can die from defoliation, but only unhealthy trees are seriously attacked by elm leaf beetles. Solutions include Spinosad, plant-oil products, and *Bacillus thuringiensis* (Bt) (see warning above). Strong populations of beneficial insects will also help.

Fire Ants: See *Ants, Fire.*

Flea Beetles: Tiny black insects that eat small round holes in foliage. Encourage biodiversity of insects, birds, plants, and small animals and spray Garrett Juice plus citrus oil. Neem sprays will also help.

Fleahoppers: These are common vegetable-garden pests that suck juices from foliage and cause a loss of leaf color, which stresses plants. Sulfur (light applications) and natural diatomaceous earth will help. Plant-oil sprays are also effective.

Fleas: Parasitic insects that attack warm-blooded animals. Fleas can be controlled with good nutrition; animals with decent diets will not have serious flea problems. Try the Muenster Natural Dog Food. See Malcolm Beck's book *Lessons in Nature* for the paunch manure story. Spray site as needed with citrus products. Apply beneficial nematodes to the soil per label directions, and bathe pets regularly with mild herbal shampoos. For dogs, make sure the shampoo contains D-limonene or citrus oil; these are too strong for cats' skin. Dusting animals with natural diatomaceous earth will also help. The best DE, because it has more than double the surface area, is DiaSource. There is anecdotal evidence that trichogramma wasps parasitize flea eggs and control the pests.

For the control of fleas and ticks, apply the following program:

1. Spray the infested site with Fire Ant Mound Drench.
2. Treat the site with beneficial nematodes. These are living organisms, so use before the expiration date on the package.
3. Dust pet sleeping quarters, if necessary, with natural diatomaceous earth. Bathe pets with mild herbal shampoos.
4. Spray the site regularly with Garrett Juice or compost tea.

Flies: Spray infested area with citrus products. Release fly parasites every two weeks when needed. Feed natural diatomaceous earth to livestock and pets at 2 percent of food volume. Flies can be repelled with fresh crushed tansy or garlic. They can also be killed with flyswatters. Hanging clear bags of water is another effective control, but they must be replaced or cleaned when dirty or dusty.

Forest Tent Caterpillars: Pest caterpillars that attack trees in the late spring or early summer. Their damage is generally more cosmetic than destructive. Release trichogramma wasps in the spring and, as a last resort, spray Bt (*Bacillus thuringiensis*) per label at dusk (see warning above). Add 1 tablespoon molasses per gallon. They can also be killed with a spray mix of orange oil, compost tea, and molasses. These caterpillars will sometimes do some damage in early spring and early summer, but if pesticides are avoided, the beneficial wasps will usually keep these guys under control. At worst they are only a temporary problem. They can also be killed with the Fire Ant Mound Drench.

Fungus Gnats: Small nuisance insects that start out in the soil of potted plants. Fungus gnats are present when the soil surface is too wet. They can also come in on bananas. They do little, if any, damage but are annoying. They can be gotten rid of by drying out the soil. Neem drench is also effective. Orange-oil spray kills them on contact. Traps can be made by putting a small amount of apple cider vinegar in a bowl with 2 drops of liquid soap. Drench the soil with neem products and allow soil to dry out between waterings. Bti (*Bacillus thuringiensis* subsp. *israelensis*) products work also but not as well as the neem drench, although the Bt product Gnatrol has proven very effective. Biocontrols include a predatory mite *(Hypoaspis miles)* that colonizes the soil and rove beetles (Staphylinidae) that voraciously eat fungus gnat larvae and eggs. Beneficial nematodes (*Steinernema feltiae*; Scanmask, NemaShield, Nemasys, or Entonem) have worked well for greenhouses or plantings inside buildings.

Grasshoppers: Pests that will eat any plant or plant part. Especially severe during drought seasons, grasshoppers are one of our most troublesome pests. Do not mow or spray the grasses and weeds under fence lines because they need the bare soil to lay their eggs. Spray Garrett Juice mixed with 2 ounces of orange oil per gallon. Add 1 quart of kaolin clay per 2 gallons of water with 1 tablespoon of Bio Wash (to clean plants). Adding garlic-pepper tea to the mix also helps. A biological bait, *Nosema locustae,* is also available and helpful with the overall program. Cover all bare soil with mulch, plant lots of different plants to establish strong biodiversity, feed the birds regularly to keep lots of them on the property, and use plants that attract birds by providing food and shelter. Heavy applications of lava sand and/or kelp meal seem to help for some reason. Dusting DE is also effective if the right DE is used and it is done during low humidity. Also, many birds such as guinea fowl, chickens, and ducks are voracious predators of grasshoppers.

Grubworms: The larvae of various beetles. The adult beetles will chew some leaves, and the grubs will eat the roots of grass and garden plants, but not all grubs are harmful; in fact, only about 10 percent of the species eat plant roots. Apply beneficial nematodes per label instructions and only use products that stimulate soil biology, such as compost, natural organic fertilizers, microbial stimulators, and rock minerals. The other 90 percent eat decaying organic matter, aerate the soil, and are beneficial. Control of the bad guys comes from being organic and having healthy soil with lots of beneficial organisms and other insects. To speed up the control, apply beneficial nematodes and small amounts of sugar or molasses.

Lacebugs: Lacebugs are flat and oval and attack various deciduous trees and broad-leaved evergreens, sucking the sap from the underside of leaves of bur oaks, azaleas, sycamores, and other plants. Black waste material will be visible. A quick solution for these pests is garlic-pepper tea and natural diatomaceous earth. Healthy biodiversity in the garden will eliminate destructive populations of this pest. Garrett Juice plus orange oil will also help. Use Bio Wash (to clean plants) as a helpful surfactant. *Note*: Research has shown that azalea lacebugs are attracted to plants that had synthetic fertilizer applied! (See C. Casey and M. Raupp, "Effect of Supplemental Nitrogen Fertilization on the Movement and Injury of Azalea Lace Bug . . .," *Journal of Environmental Horticulture* 17, no. 2 [June 1999]: 95–98.)

Leafhoppers: Small insects that do damage similar to flea beetles. Leafhoppers excrete honeydew and damage leaves by stripping them, causing stunted, dwarfed, and yellow foliage. They can be controlled with a mix of Garrett Juice and garlic tea or simply by the encouragement of diverse populations of beneficial insects.

Leafminers: Tiny insects that leave tan paths in the leaf surface and will cause brown foliage tips that often continue to spread over the entire leaf. Neem products are effective. Garrett Juice with garlic tea will help. They cause minor damage only, so treatment is rarely needed.

Leaf Skeletonizers: These insects are sawflies. They do cosmetic damage to red oak and other tree leaves; it's rarely necessary to treat. Damage is usually confined to isolated spots in the foliage. Plant-oil products are effective.

Loopers: These moth larvae eat plant foliage but are easily controlled by releasing trichogramma wasps before larvae are hatched. Apply Bt (*Bacillus thuringiensis*) product per label instructions at dusk, and add 1 tablespoon of molasses per gallon of spray. Do this only as a last resort.

Mealybugs: Mealybug adults and nymphs are sucking insects that look like cotton on plants. They suck sap from the foliage and stems and can destroy plants. Mealybugs like warm weather and also infest houseplants. Helpful controls include soap and water, predator insects, natural diatomaceous earth, and lizards. The Fire Ant Mound Drench used as a spray will also work.

For trees, spray Garrett Juice plus neem or citrus oil and apply the Sick Tree Treatment. Horticultural oil can be used in the winter months.

Mice and Rats: These rodents eat almost any green vegetation, including tubers and bulbs. Place snap traps baited with peanut butter, nut meats, or rolled oats next to walls or runways. Elimination of attractive food and water sources is an important part of rodent control. The best control of these rodents is cats. The next best is a repellent such as hot pepper spray or powder, fox urine, cinnamon, and other strong fragrances. If baits have to be used, some are safer than others. The active ingredient in Rampage is vitamin D3, which when ingested in large amounts causes hypercalcemia and quick death in rats and mice. Testing has shown this bait to be highly effective, even against anticoagulant-resistant rodent populations, yet a low hazard to people, other animals, and birds. The estimated lethal dose for a 50-pound dog is 5.9 pounds of Rampage; for a 50-pound child, 13.3 pounds. Assault is another recommended product that does not have the danger of secondary kill.

Mistletoe: This tree parasite attacks sick, stressed trees. Physical removal and soil health is the long-term control. Remove entire infested limbs where practical. Apply black pruning paint to large wounds and apply the Sick Tree Treatment. This is the only case where we recommend the use of pruning paint. In most cases, it prevents growth and hurts the plant, but in this case, it helps prevent the regrowth of the mistletoe.

Mites: Encourage frogs, birds, bats. Eliminate standing stagnant water. Spray plant products like EcoEXEMPT. See *Spider Mites* for more information.

Mole Crickets: The Basic Organic Program and resulting healthy soil will usually prevent damage from these critters. Outbreaks can be controlled with beneficial nematodes. Neem drenches can also help.

Mosquitoes: Mosquitoes are easier to control with organic techniques than with toxic chemicals. Here's the overall plan.

1. Empty standing water where possible.
2. Treat water that cannot be emptied with gambusia fish or a Bti (*Bacillus thuringiensis* subsp. *israelensis*) product such as Bactimos Briquettes or Mosquito Dunks.
3. Spray for adult mosquitoes with plant-oil products.
4. Use organic management to encourage birds, bats, dragonflies, and other beneficial insects.

5. Use skin repellents that contain natural herbs such as aloe vera, citronella, vanilla, eucalyptus, tea tree oil, and citrus oil. Do not use DEET products.
6. Bug-zapping light devices do not work! Not unless you chain your dog to the device to provide the carbon dioxide. Mosquitoes are attracted to living organisms not cold machines.
7. Apply dry granulated garlic to the problem areas.

Mosquitoes can be controlled on a large scale by spraying a D-limonene or orange-oil-based product. Adult mosquitoes can also be controlled by using any of the neem sprays or garlic sprays. The young in the water (nymphs or wigglers) can also be controlled with a gambusia (species of minnow) fish or applications of products that contain *Bacillus thuringiensis* subsp. *israelensis*. Products like Mosquito Barrier (made with garlic oil) can repel mosquitoes for several weeks. Dry granulated garlic such as Mosquito Scram broadcast on the site can repel mosquitoes for up to thirty days.

Moths: Larvae of most moths eat vegetation or fabrics and become serious pests. Spray Bt (*Bacillus thuringiensis*) per label instructions at dusk (see warning above). Release ladybugs and green lacewings every two weeks until natural control exists. Indoors, use repellents of cedar or wormwood. Bay leaves also help. Release trichogramma wasps in the spring.

Nematodes: Over fourteen hundred species have been identified, of which only twenty or so eat roots and are pests; the rest are good guys. Many nematodes are beneficial, but there are those that will attack ornamental trees, garden plants, and lawn grass. Root knot nematodes are plant-damaging soil organisms. Some nematodes attack plant roots and greatly reduce production. Nematode invasion symptoms include wilt, reduced growth, and lack of vigor. Control measures involve crop rotation, enriching the soil with humus, and planting pest-free stock. Liquid biostimulants help.

A few types of nematodes are barely visible to the naked eye. Larvae, which hatch from eggs, molt several times before maturing into adults. Troublesome nematodes puncture plant cell walls, inject saliva, and suck out the cell's contents. Some species move from plant to plant in water or on garden tools; others attach themselves permanently to one root. Some nematodes cause excessive branching of roots, rotted roots, or enlarged lumps on roots. Other nematodes simply reduce the size of root systems.

Carrot and other root crop plants attacked by nematodes will be stunted with yellowed leaves, and roots may be distorted or have swollen areas. Legumes are supposed to have swellings on their roots that are caused by nitrogen-fixing bacterial. The difference is that the bacteria nodules are attached to the outside of the root. Nematodes make it swell from the inside.

Orange peelings tilled into the soil are often the most effective control. Till or fork citrus pulp into soil prior to planting or gently work into the root zone of existing plants. Stimulate soil biology with compost, natural organic fertilizers, and microbe stimulators. Cedar flakes used in the mulch of outdoor gardens and as the floor material in greenhouses repel nematodes very effectively. Several species of marigolds have been found to help control nematodes and some weeds. *Catharanthus roseus* (formerly *Vinca rosea*), periwinkle or vinca, is a perennial that has also been found to kill harmful nematodes.

Peach Tree Borers: Adults are 1-inch clear-wing narrow blue-black wasps that look like moths with a wide yellow-orange belt. Eggs are brown or gray and laid in the bark or in the soil near the base of fruit trees. Larvae are 1-inch white or pale yellow caterpillars with brown heads that burrow into the trunk near the ground line. One generation per year is produced. Larvae hibernate in burrows in trees or in soil; pupates, in burrows or in brown cocoons in the soil. Host plants include peach, plum, apricot, and cherry trees. Borers (larvae) chew inner bark of lower tree trunk, and a mass of gummy sawdust appears at base of trees. They are also vectors of wilt fungi and other diseases. Natural control includes healthy soil, compost, mulch over root system. Organic control includes tobacco dust (snuff) or sticky tape around base of tree, the Organic Pecan and Fruit Tree Program (see Appendix 1), and the Sick Tree Treatment.

Pecan Nut Casebearers: The larvae of these small moths damage pecan nut production by feeding on the small nutlets as they develop. To control, release trichogramma wasps every two weeks during the spring; first release should be made at leaf emergence. A little damage from this insect is actually beneficial, as it reduces heavy nut production.

Pillbugs; Sowbugs: Pillbugs and sowbugs, or roly-poly bugs, are crustaceans related to shrimp, crabs, and crawfish. They are found in damp places and feed on organic matter, but when abundant, will also eat plants. A mix of cedar flakes, hot pepper, and natural diatomaceous earth gives effective control. Mulch using shredded cedar. Spray plant oil or Spinosad.

Pine Beetles: The southern pine beetle (*Dendroctonus frontalis* Zimmermann) is one of pine's most destructive insect enemies in the southern United States, Mexico, and Central America. Because populations build rapidly to outbreak proportions and large numbers of trees are killed, this insect generates considerable concern among managers of southern pine forests. The solution is healthy soil and healthy trees. Orange-oil sprays can be used to treat infestations.

Pine Tip Moth: Small moth larvae that attack the terminal growth of pine. It is an easy pest to control with organic techniques. Release trichogramma wasps and spray Garrett Juice plus garlic. Only add the garlic to the spray if needed.

Plum Curculio: A common insect pest of fruit trees. Plant adapted varieties and use the Organic Pecan and Fruit Tree Program. Make sure the root flares of the trees are exposed. Regular spraying of garlic tea is one the best organic preventatives. Biodiversity is critical for control of this pest. Spray foliage bi-weekly with Garrett Juice plus garlic tea.

Rats: See *Mice and Rats.*

Red Ants: See *Ants, Harvester.*

Roaches: See *Cockroaches.*

Scale: Scale insects attach to stems, branches, and trunks and suck sap from the plants. Controls include horticultural oil and plant-oil products. Use Bio Wash (to clean plants) and water with a mild orange spray on interior plants. The black, scale-eating ladybug feeds on scale insects outdoors. Expose root flares of trees and apply the entire Sick Tree Treatment.

Sapsuckers: These beautiful birds attack and severely damage stressed trees. Sick trees build up complex carbohydrates or sugars as a defense. The birds can detect that condition and go after the sweet sap. Spray the trunks with garlic-pepper tea or an orange-oil mix. Apply the Sick Tree Treatment and treat wounds with Tree Trunk Goop.

Slugs and Snails: Slugs and snails must be kept moist at all times, so they will go anywhere there is moisture. Effective controls include garlic-pepper tea, natural diatomaceous earth, and wood ashes. They can also be repelled with a mix of natural diatomaceous earth, hot pepper, and cedar flakes. The commercial product Sluggo is also effective, and orange-oil sprays will also kill them. The larva of a parasitic nematode (*Phasmarhabditis hermaphrodita*) enters the slug and releases a bacterium that kills the slug. Used where appropriate, ducks love to eat slugs and snails and are extremely effective at eliminating them; however, too much duck traffic can cause compaction. Some other slug predators are ground beetles, rove beetles, firefly larvae, snakes, turtles, frogs and toads, salamanders, birds, and small mammals.

Sowbugs: See *Pillbugs; Sowbugs.*

Spider Mites: Red spider mites or just spider mites are very small spider kinfolks that feed on garden plants and ornamental trees that aren't able to absorb water properly. Too much or too little water can be the culprit. You probably will not see the mites at first, but you will notice the webbing that accompanies them. The best control is proper watering and the use and protection of beneficial insects such as green lacewings. Controls also include spraying Garrett Juice plus garlic and extra seaweed. Seaweed spray by itself also works. Strong blasts of water or Bio Wash (to clean plants) are good for small infestations. Predatory mites are also effective.

Spiders: Physically remove—no need to control spiders except black widow and brown recluse. Encourage them, as most are beneficial in controlling pest insects. Spray if needed with citrus or Spinosad products. Release and protect gecko lizards.

Squash Bugs: Light gray true bugs are difficult-to-control insects that attack squash, cucumbers, pumpkins, and other cucurbits. Control by smashing the eggs, dusting the adults with natural diatomaceous earth, and planting lemon balm in between plants. Dusting young plants regularly with cheap self-rising flour will also help. Treat soil with beneficial nematodes and spray with Garrett Juice plus orange oil. Planting a larger number of squash plants also seems to help. Hand remove and destroy eggs from the back side of leaves. Gardeners have reported successful control by grinding bay leaves and working the material into beds before planting.

Squash Vine Borers: The squash vine borers are insects whose larvae are worms that bore into the base stem of squash, cucumbers, melons, gourds, and pumpkins. Cut the stem open, remove the worms, and cover the wounded area with Tree Trunk Goop. Another way is to inject Bt into the base of the stem with a syringe. Spraying young plants with Bt will also help. Treat soil with beneficial nematodes. Adults are pretty red-and-black moths that look like wasps. Plant a second crop after two weeks because the pests only produce one generation per year.

Stink Bugs: Stink bugs punch holes in flowers, fruits, and leaves and cause rotted spots. Some stink bugs, on the other hand, are beneficial. The pest ones can be controlled with a spray of Garrett Juice plus orange oil or a dusting of natural diatomaceous earth. EcoEXEMPT also works.

Termites: Treat all exposed wood with borate products such as Tim-Bor or Bora-Care or with hot pepper. Inject these same products by foam into the walls. Use 00 sandblasting sand (also sold as 16-grit sand) as a physical barrier around pipes and conduits in slabs, against the edge of slabs, and on both sides of beams. New construction can use it under slabs. Treat the soil around the structure with beneficial nematodes. Ignore those who recommend removing the mulch from around the house. Dust a mix of 90 percent natural diatomaceous earth and 10 percent boric acid into the wall cavities before closing them in. This technique sets up a permanent pest control.

Thrips: Thrips attack buds and tight-petaled flowers such as roses, mums, and peonies. Thrips are not visible to the naked eye but will rasp the plant tissue and drain the sap. Heavy infestations can kill plants. Thrips are general eaters and will attack flowers or field crops. Controls include spraying Garrett Juice plus garlic and releasing green lacewings. Biweekly spraying of Garrett Juice is all that is usually needed long term. Apply beneficial nematodes to the soil before bud break in the early spring.

Ticks: Ticks are difficult to control, but dusting with natural diatomaceous earth and using the flea program will control these pests. Bathing the pets regularly with herbal ingredients will help considerably. Spray an essential-oil product such as EcoEXEMPT about 8 feet up on outdoor walls and the trunks of trees and shrubs because that is where ticks hang out waiting for a blood meal to come strolling by. Apply beneficial nematodes in late spring to kill the young ticks in the ground.

Note: One benefit of leaving a few fire ant mounds around is that fire ants love to eat ticks, fleas, chiggers, and termites and are effective at removing them from a property.

Tobacco (and Tomato) Hornworms: Large green caterpillars that love tomatoes. Spray Bt (*Bacillus thuringiensis*) per label instructions plus molasses at 1 ounce per gallon of spray at dusk (see warning above). Release beneficial wasps every two weeks. Handpick as they start to feed.

Tomato Pinworms: Minor pest problem. Spray every two weeks with Garrett Juice, garlic, garlic-pepper tea, or citrus oil as needed. Release green lacewings and trichogramma wasps. Apply beneficial nematodes.

Treehoppers: Pretty little insects that have a triangular shape. Treat the soil with horticultural cornmeal. Use the Organic Pecan and Fruit Tree Program. Spray Garrett Juice plus garlic or garlic-pepper tea for serious infestations; spray every two weeks or as needed. They are seldom a serious problem.

Wasps: Nests can be moved to a new location and nailed in place after spraying wasps with water. Do not attempt if allergic to wasps. Mild citrus sprays will repel them, but do not spray directly on the wasps. Protect wasps if possible, as they are highly beneficial, but unwanted insects can be repelled by painting porch and overhang ceilings with "Haint Blue" paint. Painting trim and other surfaces will help as well.

Webworms: Webworms are the larvae of moths. Release trichogramma wasps, spray with Garrett Juice plus garlic. Add neem or citrus oil for strong infestations. Use Bt (*Bacillus thuringiensis*) as a last resort. Spray at dusk with 1 tablespoon molasses added per gallon. The best control is to protect the paper wasps and other native wasps.

Whiteflies: Small white insect pests that resemble little white moths. Whiteflies are extremely hard to control with chemicals and will suck the juices from several kinds of plants. Spray as needed with garlic-pepper tea at ¼ cup per gallon, Garrett Juice plus garlic or orange oil, seaweed plus garlic-pepper tea, or Spinosad. Release green lacewings as needed. Old university research has shown that a phosphorus deficiency will cause an infestation of this pest. Gardeners and growers using rock phosphate liberally rarely have whitefly problems.

Structural Pest Control

Organic pest control for structures is a frustrating issue because the structural pest control board (for example, the one in Texas) does not allow pest control companies to use beneficial nematodes for the control of termites, even though it is an excellent control. The alleged reason is that they are not labeled for the control of termites. So people have to take the responsibility to use them without professional help. A barrier of 00 sandblasting sand along the edges of and under slabs and around piers is also not accepted as an approved tool in some states, although it works beautifully.

The accepted products that are used and should make up the primary part of termite control are the boric acid and natural diatomaceous earth products, which can be dusted into existing wall cavities. Borate products can be painted on new construction—the studs, flooring, and sills—and give very

rganic programs can be successful at any scale and are better than the toxic chemical approach in every way—ncluding the economics of the project. Photograph by Howard Garrett.

oil—before and after—organic amendments and in the absence of synthetic fertilizer and toxic pesticides. hotograph by John Ferguson.

Pine tree seedlings in growth chamber. The seedling on the right was inoculated with mycorrhizal fungi; the one on the left was not. Photograph by Mike Amaranthus.

Potatoes without (left) and with (right) a biologically active root zone. Photograph by Mike Amaranthus.

Rock minerals and sugars—lava sand on the left and dry molasses on the right—are as important to the organic soil system as compost and organic fertilizers. Photograph by Howard Garrett.

Nursery-grown transplant showing all the wrong ways to grow plants—in peat moss with perlite or foam pellets, which leads to a dead, dry, and unhealthy plant showing no biological activity. Photograph by Howard Garrett.

Tree planting—the right way. Photograph by Howard Garrett.

Tree planting—the wrong way. Photograph by Howard Garrett.

Bare-root planting techniques. Photograph by Howard Garrett.

Tomatoes growing in soil not treated with mycorrhizal fungi (on left) and treated (on right). Photograph by Mike Amaranthus.

Synthetic "weed and feed" fertilizers should be avoided. Photograph by Howard Garrett.

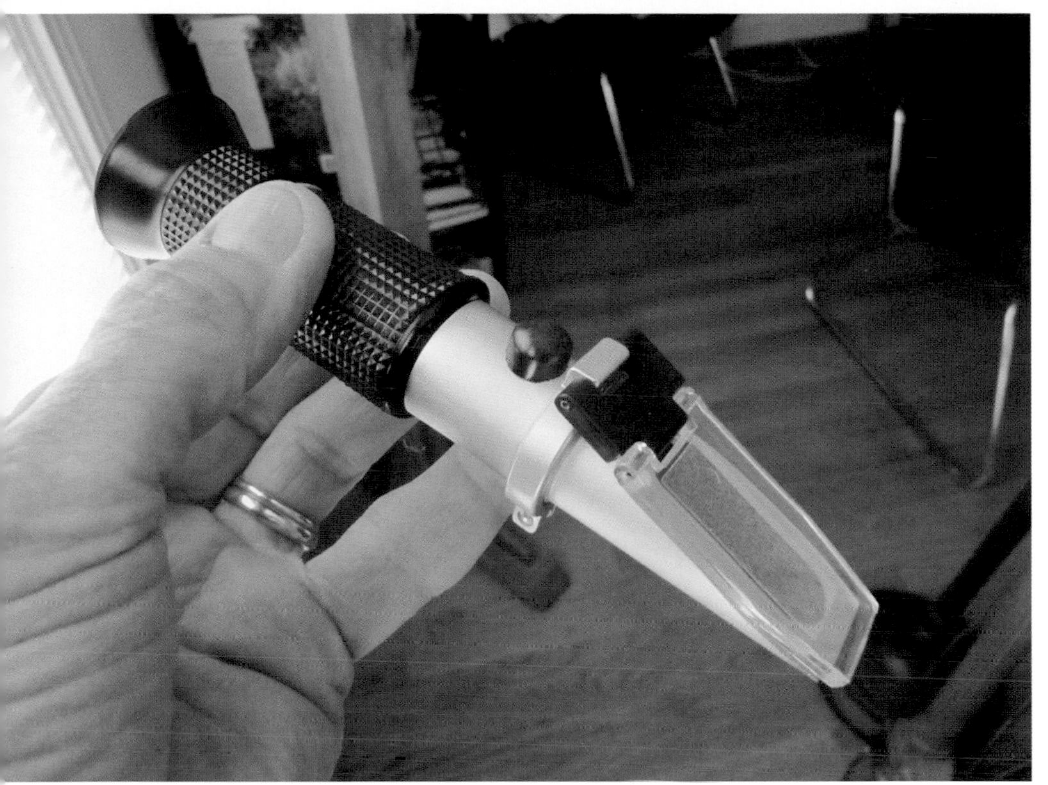

Refractometer used to measure (in degrees Brix) the sugar in plant tissue. An excellent tool to learn what nutrients are getting from the soil into plants. Photograph by Howard Garrett.

Trichogramma wasp—one of the most efficient and cost-effective beneficial insects that can be purchased (on ards, as here, or strips) and used to control many species of lepidopterous insect pest larvae. Photo shows moth eggs ontaining the entire trichogramma wasp life cycle. Photograph by Howard Garrett.

Hover fly—an important insect pollinator whose larvae are effective predators helping with pest control. It is often misidentified as a wasp. Photograph by Howard Garrett.

Honeybees being managed by Dallas beekeeper Brandon Pollard, who understands the dangers to bees from toxic pesticides and thus uses natural organic techniques and products. Photograph by Howard Garrett.

Properly made compost with beneficial fungi visible. Photograph by Howard Garrett.

Shredded native tree trimmings mulch—the best mulch in many ways. Photograph by Howard Garrett.

Trunk flares exposed—one of the most important goals of proper tree-planting techniques. Photograph by Howard Garrett.

Commercial organic project—Texas Discovery Gardens at Fair Park, the State Fair grounds in Dallas, "the first public garden in the state of Texas to be certified 100% organic by the Texas Organic Research Center" (http://texasdiscoverygardens.org/the_gardens.php). Photograph by Howard Garrett.

Bare soil in the root zones of trees—something that should never be allowed. Photograph by Howard Garrett.

Tierra Verde Golf Club in Arlington, Texas—significant cost savings resulted from a switch to a 100 percent organic fertilization program. Photograph by Howard Garrett.

Test spot treated with compost on a sports field in Houston, Texas. Photograph by John Ferguson.

Satellite image of damage to Gulf of Mexico caused by runoff of synthetic fertilizers from agricultural crops. Photograph courtesy of USDA.

)rganic trials at the USDA fields in the Rio Grande Valley in Texas. These cotton plants are being grown under he traditional chemical program. Photograph by Howard Garrett.

Cotton plants showing greater growth and more bolls per plant in the fields in Joe Bradford's organic trials. This ield uses organic fertilization and no toxic pesticides. Photograph by Howard Garrett.

Corn-growing research by Mike Amaranthus. Tote on the left used organic fertilizer and mycorrhizal fungi. Tote on the right used recommended synthetic fertilizer. Photograph by Mike Amaranthus.

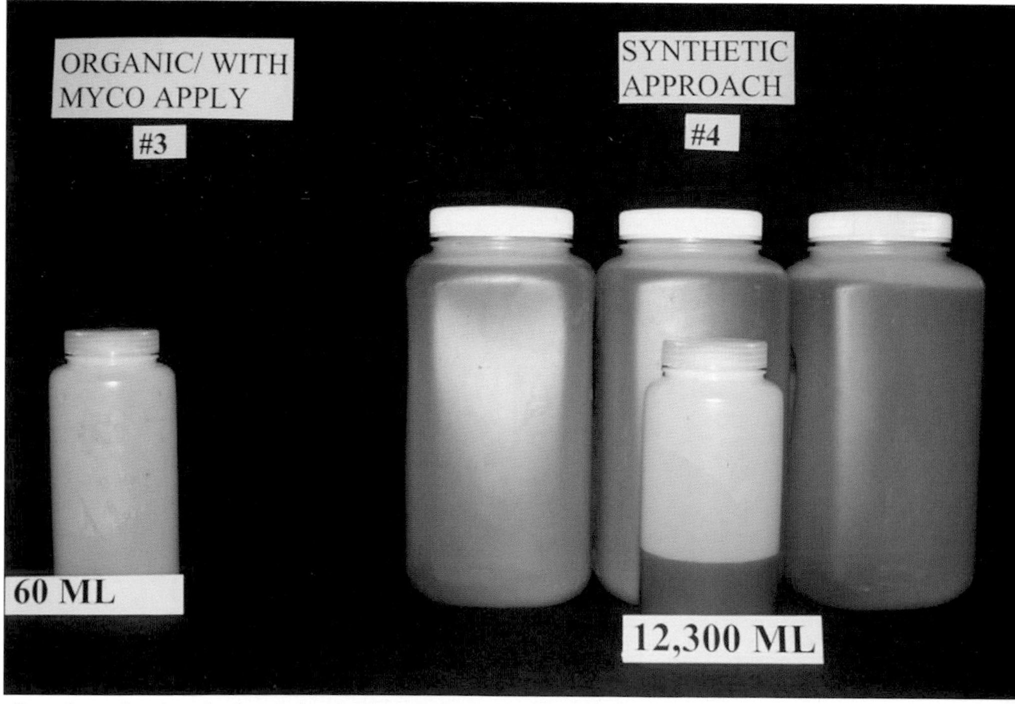

Containers showing the dramatic reduction of water runoff and lost nutrients in the organically fertilized plants (left) compared with the synthetically fertilized plants (right). Photograph by Mike Amaranthus.

long-lasting protection against carpenter ants and termites. Available products include Tim-bor and Bora-Care. Beneficial nematodes applied to the soil around buildings help greatly with termites and many other pests that live at least part-time in the soil. Orange-oil and D-limonene products kill and repel carpenter ants and other structural pests. The products that contain various plant oils are excellent for controlling most pest insects but are toxic to mammals. One of the leading companies is EcoEXEMPT.

Honeybees and Other Pollinators

Honeybees being managed by Dallas beekeeper Brandon Pollard, who understands the dangers to bees from toxic pesticides and thus uses natural organic techniques and products. Photograph by Howard Garrett.

One of the disturbing things about farming in general is the continued loss of our pollinators. Not only honeybees and bumblebees but all the beneficial insects feed on nectar and pollen and help us with pollination. The culprits of the mysterious bee colony collapse disorder (CCD) are many. One is the varroa mite and the other the tracheal mite. Researchers reported that it is due to two common infections, one viral (*Nosema ceranae*) and one virus (IAPV), working together. The mites and these two diseases result from stress caused by pesticides and other environmental pressures. To our dismay, we found that it is common in the beekeeping industry to treat the beehives with toxic chemical pesticides. That simply has to be stopped. Some of the beekeepers have become as big a problem as the farmers and ranchers who are still using the chemical poisons and killing insects in a broad-spectrum way. Spraying for boll weevils has probably killed off a lot of the bees. There is nothing about the boll weevil eradication program that makes any sense at all, and the secondary killing of bees is another good reason for the program to be abandoned.

The question of how to control the mites remains. Like all pest problems, this one seems to be due to a reduced immune system in the bees. We need to figure out how to change that. Beehives are sometimes fed high-fructose corn syrup (HFCS), which often contains the deadly HMF contamination. Leaving some of the bees' own honey may go a long way in helping to solve the problem. Another possibility that has already shown some positive results is painting the beehives a darker color, thus increasing the temperature in the beehives. The bees flex their wing muscles, raising the temperature in their bodies, and that apparently helps to get rid of the mites.

Mite control with essential oils from peppermint, spearmint, thyme, and lemongrass, for example, seems to have promise. One benefit of this treatment is that there is a wide range of effective and potentially effective essential oils, which makes it less likely for the mites to develop immunity. Feeding bees syrup containing these oils enables the nurse bees to feed the essential oil to the larvae. Growing herbs in the beehive areas is a good idea. Bees will get the oil from pollen and additional fragrance from the flowers.

Another control that seems to be very effective is using screened bottom boards and dusting powdered sugar on the bees. It works by stimulating the bees to groom themselves more thoroughly. This knocks the mites off, and they fall through the screen bottom of the hives.

Plant pollination from bees is very, very important, as is having high-quality local honey. Here are some of the best plants for attracting honeybees and other pollinators.

Trees and Shrubs

Abelia
Ash
Blueberry
Butterfly Bush
Buttonbush
Catalpa
Crabapple
Golden Rain Tree
Hawthorn
Holly
Honeysuckle
Indian Hawthorn
Indigo
Japanese Maple
Lilac
Linden
Magnolia
Maple
Mexican Plum
Mountain Ash
Oak
Pear
Persimmon
Poplar
Privet
Redbud
Sweet Mock Orange
Sycamore
Tulip Tree
Witch Hazel
Willow

Annuals and Perennials

Ajuga
Almond Verbena
Aster
Basil
Bee Balm
Borage
Buttercup
Catnip
Chives
Clematis
Clover
Columbine
Comfrey
Coriander/Cilantro
Cosmos
Crocus
Daffodil
Dahlia
Echinacea
Fennel
Foxglove
Geranium
Germander
Globe Thistle
Goldenrod
Grape Hyacinth
Greg's Mistflower
Hollyhock
Hyacinth
Joe-Pye Weed
Lavender
Lemon Balm
Lovage
Marigold
Marjoram
Milkweed
Mint
Obedient Plant
Peppermint
Poppy
Rose
Rosemary
Sage
Sedum
Snowdrop
Spider Plant
Strawberry
Sunflower
Tansy
Thistle, Globe
Thyme
Yarrow
Zinnia

Diseases

Disease control in an organic program is easy. Increased resistance to most diseases results as a nice side benefit from the Basic Organic Program. All organic products help control disease to some degree. When soil and plants are healthy, there is a never-ending microscopic war being waged between the good and the bad microorganisms, and the good guys win. Disease problems are simply situations in which the microorganisms have gotten out of balance.

Drainage is a key ingredient for the prevention of diseases. Beds or tree pits that hold water and don't drain properly are the ideal breeding place for many disease organisms.

As with insects, spraying for diseases is only treating the symptoms, not the major problems—plus the toxic sprays kill more beneficials than the targeted pests or diseases. The primary cause of problems is usually related to the soil and the root system. Therefore, it is critical to improve drainage, increase air circulation, add quality composted products, and stimulate and protect the living organisms in the soil. When oxygen levels in the soil drop below 6 ppm, soil diseases start actively growing (often due to compaction, overwatering, hardpan, salts from chemicals, plastic mulch, etc.).

Anthracnose: A fungal problem in sycamore trees, beans, and ornamentals that causes the foliage to turn a tan color overnight. Sycamore leaves turn brown, remaining on the tree. Not normally fatal. Spray Garrett Juice plus garlic tea on emerging new foliage in early spring. Apply the Sick Tree Treatment. Sprays of potassium bicarbonate, garlic, and neem are also effective. Control is difficult other than by avoiding the use of susceptible plants. Also called bird's-eye spot, anthracnose causes small dead spots with a raised border, a sunken center, and concentric rings of pink and brown. Symptoms of bean anthracnose manifest on the pods as circular black sunken spots that may ooze pink slime and develop red borders as they age. Thoroughly compost infected plants. Treat the soil with cornmeal and use the overall Sick Tree Treatment. Spray infected plants with 2 percent hydrogen peroxide or Bio Wash (to clean plants).

Asparagus Rust: Fungal disease that appears as a browning or reddening of the foliage and a release of rusty, powdery spores. It can overwinter and infect new shoots as they emerge the following spring. Rust is moved from plant to plant by wind. To control, spread out plants to allow air circulation and plant resistant varieties. Remove infected plants and thoroughly compost them. Spray plant with Bio Wash (to clean).

Bacteria: Single-cell microorganisms that reproduce by simple cell division. Rots cause decay of leaves, stems, branches, and tubers. Vascular system blockage causes wilting. Galls result from an overgrowth of the affected cells. Bacterial problems are encouraged by poor drainage, wet soil, high humidity, and high temperatures. Feeding plants with slow-release natural organic fertilizers will help prevent bacterial infection. Use disease-free seed and resistant varieties. Remove infected plants promptly and compost. You need a microscope to see the actual bacteria, but the symptoms they cause are easy to see with the naked eye. Most bacteria break down dead organic matter and are beneficial. A few, however, cause plant diseases. Bacteria usually reproduce by splitting in half. They are spread by wind, water, insects, garden tools, and gardeners' hands. Bacterial diseases are more difficult to control than fungal diseases.

Bacterial Blight: A bacterial disease that causes dark green water spots that turn brown, leaving a hole in the leaves of tomatoes, plums, and several ornamental plants. Controls include healthy soil, garlic tea, hydrogen peroxide, and Bio Wash (to clean plants).

Bacterial Leaf Scorch: A fatal disease of sycamores and other plants. It causes a browning between the veins of the leaves. It kills limbs from the tips and progresses down the branch quickly. Often incorrectly diagnosed as anthracnose. Treat with Garrett Juice and Bio Wash (to clean plants).

Black Spot: Common name of fungal leaf spot. Black spot attacks the foliage of plants such as roses. There is usually a yellow halo around the dark spot. Entire leaves then turn yellow and ultimately die. Best controls include selection of resistant plants and Garrett Juice plus garlic tea and skim milk. Keep bare soil mulched. Apply cornmeal to the soil or cornmeal juice to the foliage. Bio Wash (to clean plants) is the best commercial product. It works great!

Blight: When plants suffer from blight, leaves and infected branches suddenly wither, stop growing, die, and may rot. Drench the soil with neem.

Brown Patch: Cool-weather fungal disease of St. Augustine grass. Brown leaves pull loose easily from the runners. Small spots in lawn grow into large circles that look bad and weaken the turf but rarely kill the grass. Soil health, drainage, and low-nitrogen input are the best preventatives. Treat diseased turf with cornmeal at 10–20 lbs. per 1,000 square feet or dry granulated garlic at 2 lbs. per 1,000 square feet. Spray Garrett Juice with Bio Wash added. Good-quality compost can also be used as both a preventative and a curative for brown patch.

Brown Rot: A fungal disease that is common in fruit trees not using an organic program. It may cause serious damage to stone fruits during wet seasons. Prolonged wet weather during bloom may result in extensive blossom infection. Early infections appear as a blossom blight or twig canker. Later infections appear as a rot of ripening fruit on the tree and in storage. Spring infections arise from mummified fruit of the previous season that remains attached to the tree or has fallen to the ground.

The solution is the organic program and preventative spraying with garlic-pepper tea, neem, hydrogen peroxide, or Bio Wash (to clean plants). Aerated compost tea is also very effective. Make sure root flares are visible and not covered with mulch or soil.

Canker: A stress-related disease of trees and shrubs that causes decay of the bark and wood. Cankers have to start with a wound through the bark. Healthy soil and plants are the best solution. Use Tree Trunk Goop on the injured spots, improve the environmental conditions, and apply the Sick Tree Treatment. Synthetic fungicides do not work on this disease. Hypoxolyn canker is a common disease of certain oaks, such as stressed post oaks, especially after droughts or long rainy seasons. The brown spores rub off easily, and the bark sloughs off the trunks. No treatment is necessary other than improving the immune system of the tree. This is the number-one disease seen on post oaks. The stress leading to this disease is often caused by herbicide treatment to the root zone. Look for cracks, sunken areas, or raised areas of dead or abnormal tissue on woody stems. Cankers ooze sometimes and can girdle shoots or trunks, causing the plant above the canker to wilt and die. Blights and canker diebacks look quite similar. Cold-injury symptoms can look like or lead to the development of cankers.

Cedar-apple Rust: Cedar-apple rust is just one of several similar fungal diseases that could broadly be classified as juniper-rosaceous rusts. All have very similar disease cycles but differ by which juniper and rosaceous species they infect. The fungus spends part of its life cycle on a juniper host and part on a host in the rose family. It requires both hosts to complete its life cycle. Cedar-apple rust is caused by the fungus known as *Gymnosporangium juniperi-virginianae.* Two other common juniper-rosaceous rusts are hawthorn rust and quince rust, although there are many more. Examples of juniper hosts include eastern red cedar, southern red cedar, Rocky Mountain juniper, some prostrate junipers, and Chinese juniper. Examples of rosaceous hosts are apple, crabapple, hawthorn, quince, serviceberry, and pear. Some commercial apple varieties are highly susceptible to cedar-apple rust with both direct fruit infection and defoliation of infected leaves. Apply the Sick Tree Treatment and spray Bio Wash (to clean plants).

Chlorosis: A condition caused in various plants by trace mineral deficiency. Iron scarcity is usually blamed, but the cause can be the lack of several trace minerals or magnesium or soils being out of balance. To cure, improve trace mineral availability by applying greensand, humate, Flora-Stim, or natural organic fertilizer and by foliar feeding. After minerals are applied, spread a natural mulch for continued control.

Club Root: Fungal disease that attacks vegetables and flowers in the cabbage family. Plants wilt during the heat of the day. Older leaves turn yellow and drop, and roots are distorted and swollen. To control, select resistant cultivars, rotate related crops, and thoroughly compost sick plants. Apply liquid biological activators and cornmeal.

Construction Damage: Construction activity causes compaction of the soil, which squeezes out oxygen and kills beneficial microbes and root hairs of trees and other plants. Prevent it—don't allow it—use physical barriers. Build strong fences so contractors cannot access the root zone. Use coarse mulch on all work areas and areas of vehicle access (roads and parking). Three inches of coarse, shredded native tree-trimming mulch is as effective as ¾-inch plywood at preventing compaction. The mulch also prevents tracking of mud and dirt into new construction.

Cotton Root Rot: A fungal disease common in alkaline soils that attacks poorly adapted plants. The best preventative is healthy soil with a balance of nutrients and soil biology. Treat the soil with cornmeal at 10–20 pounds per 1,000 square feet. Adding sulfur to the soil at 5 pounds per 1,000 square feet annually will help. Do annual soil tests, and stop adding sulfur when enough is in the soil. In no-till agriculture, the organic content comes up. This is a human-made disease; chemicals destroyed the humus and protective living organisms.

Crown Gall: A bacterial disease that infects and kills grapes, roses, fruit trees, shade trees, flowers, and vegetables. Galls are rounded with rough surfaces and are made up of corky tissue. They can appear on stems near the soil line, on graft unions, and on roots or branches. Don't buy suspicious-looking plants and don't replant in an area where you have had crown gall. Avoid wounding stems, and disinfect tools with hydrogen peroxide between plants when pruning. Destroy infected plants or prune away galls.

Crown Rot: Caused by *Phytophthora* pathogens (oomycetes), which attack the roots and, most notably, the crown of plants such as African violets. Plants are most susceptible when allowed to sit in soil that is heavy and soggy. Crown rot is often fatal. The best control is prevention.

Symptoms: Crown is mushy and may appear translucent brown with a soft, jellylike consistency. Its brown or black color is not to be confused with the web left behind by spider mites, which clings to the flowers and plant hairs of the leaves. Leaves darken and appear brown or black in color. Leaves wilt and may appear translucent. Controls include cinnamon, hydrogen peroxide, cornmeal tea, and Bio Wash (to clean plants).

Curly Top: A viral disease that attacks vegetable crops such as tomatoes. To prevent, control the aphids that spread the disease. Use shiny material such as silver Mylar under plants. Drench soil with neem product. Spray plants with 3 percent hydrogen peroxide—it really works but can burn plants during hot weather. Spray during the coolest part of the day.

Damping-Off: A fungal disease of emerging seedlings that causes tiny plants to fall over as if severed at the ground line. Avoid by using living organic (not sterilized) potting soil and by placing rock phosphate on the surface of planting media. Treat the soil with cornmeal and spray with Bio Wash (to clean plants). This disease is caused by a number of fungi, mainly *Pythium* and *Rhizoctonia,* and the pathogen *Phytophthora.* The symptoms include decay of seeds prior to germination, rot of seedlings before emergence from the root medium, and development of stem rot at the soil line after emergence. The collapse of seedlings at or just below soil level is very common in synthetic programs where unhealthy potting soil is used. It is especially bad when there is overwatering and/or poor drainage. Avoid this disease by using colloidal phosphate or cornmeal on the surface of planting media. Keep soil moist but not waterlogged. Provide good air movement and always use living organic material such as earthworm castings or compost in the seed beds. Wait until soil is warm enough for the specific plant before seeding.

Dollar Spot: A fungal disease (*Sclerotinia homeocarpa*) that forms small brown spots on golf green grasses and other Bermuda turfgrasses. It can be controlled with various cornmeal products. Tierra Verde Golf Club in Arlington, Texas, controlled it with corn gluten meal at rates used for fertilizing (20 pounds per 1,000 square feet). It attacks most turfgrasses grown in the South. Bentgrass, hybrid Bermuda grasses, and zoysia are most susceptible to this disease. It occurs from spring through fall, and is most active during moist periods of warm days (70°F–85°F) and cool nights (60°F) in the spring, early summer, and fall. It is spread by water, mowers, other equipment, and foot traffic. To prevent dollar spot, use proper cultural practices and the organic program to promote healthy turf. Avoid light, frequent waterings. To cure dollar spot infection, apply cornmeal or garlic in dry or liquid forms as needed. Applications are most critical during moist weather in the spring, early summer, and fall when temperatures are between 70°F and 80°F.

Downy Mildew: Fungal disease that attacks fruits, vegetables, flowers, and grasses. The primary symptom is a white to purple downy growth, usually on the underside of leaves and along stems, which turns black with age. Upper leaf surfaces have a pale color. It can overwinter on infected plant parts and remains viable in the soil for several years. It is spread by wind and rain and in seeds. To control, thoroughly compost sick plants and apply cornmeal to the soil surface.

Early Blight: A fungal disease that infects ornamental plants, vegetables (tomatoes, potatoes, and peppers), fruit trees, and shade trees. Brown to black spots form and enlarge on lower leaves, developing concentric rings like a target. Heavily infected leaves dry up and die as spots grow together. Targetlike sunken spots will sometimes develop on tomato branches and stems. Both fruits and tubers can also develop dark, sunken spots. Spores are carried by the air and are a common cause of hay fever allergies. Control this disease by planting resistant cultivars and soaking seed in a disinfecting solution such as a hydrogen peroxide mixture before planting. Spray plants with compost tea and treat soil with cornmeal products.

Entomosporium Leaf Spot: A fungal disease of photinias, hawthorns, and other related plants. It primarily hits large monoculture plantings and is most active in spring and fall. It can be controlled by improving soil conditions and avoiding susceptible plants. Use the Sick Tree Treatment and try to avoid watering the foliage. Although it shows up on the foliage as round dark purple spots, it is really a disease of the root system. Baking soda or potassium bicarbonate spray will stop the spotting on the leaves if caught early. Improving the health of the root system with aeration, compost, and rock powder is the long-range cure. Products for the soil containing alfalfa will also help. Use the Sick Tree Treatment for ultimate control. Drenching the soil with mycorrhizal-laced compost tea products is the most effective solution if only one treatment is to be done.

Fire Blight: A bacterial disease of plants in the rose family in which blossoms, new shoots, twigs, and limbs die back as though they have been burned. Leaves usually remain attached but often turn black or dark brown. Prune back to healthy tissue and disinfect pruning tools with a 3–5 percent solution of hydrogen peroxide. Spray plants at first sign of disease with Garrett Juice plus garlic and/or neem. Consan 20 and agricultural streptomycin are also effective controls. Kocide 101 is a copper-based fungicide often recommended; some consider this organic, but we don't. The best recommendation is to spray Garrett Juice plus garlic, treat the soil with horticultural cornmeal, apply the Sick Tree Treatment, and reduce the nitrogen fertilizer. High-nitrogen synthetic fertilizers are the primary cause of this disease.

Fruit Rot: Grapes infected with fruit rot turn brown (so why don't they call it brown rot?), then harden into small black mummified berries. Brown rot of stone fruits causes whole fruit to turn brown and soft. Control fruit rots by planting resistant cultivars, removing and destroying infected fruit, and pruning to increase air movement. Spraying with seaweed, garlic, compost tea, and sulfur throughout the season also helps. The Sick Tree Treatment will help prevent or cure this disease.

Fungi: Fungi are microscopic primitive plants that lack chlorophyll and produce tiny spores that are spread by wind, water, insects, and gardeners. Spores germinate to form mycelia, which are the fruiting bodies. Mycelia rarely survive winter, but spores easily survive from season to season. Many fungi live on and help decompose dead organic materials. These beneficial fungi are allies in the organic garden. Parasitic fungi, on the other hand, are leading causes of plant diseases. Most disease fungi live on a host plant, in the soil, or on organic matter at various times during their life cycle. Some attack only one species of plants, while others attack a wide array of plants. Fungi are the microbes that are the easiest to see with the naked eye. They are sometimes seen as round or free-form shapes on plant foliage. Downy mildew grows from within a plant and sends out branches through the stomata to create pale patches on leaves. Powdery mildew lives on the leaf surface and sends hollow tubes into the plants. Rust fungi are named for the reddish color of their pustules. Leaf spot fungi cause yellow-green spots with black exterior rings. Soil-inhabiting fungi cause damping-off, which kills small seedlings. Many fungi are encouraged by constant moisture on foliage. Various natural treatments that stimulate *Trichoderma virens* work, such as cornmeal, cornmeal juice, garlic tea, etc.

Fungal Leaf Spot: See *Entomosporium Leaf Spot.*

Fusarium and Verticillium Wilt: Fungal wilts that attack a wide range of flowers, vegetables, fruits, and ornamentals. Plants wilt and usually turn yellow. To control, plant resistant cultivars or treat with cornmeal products. Remove affected branches and thoroughly compost them or the entire plants.

Galls: Swollen masses of abnormal plant tissue caused by fungi, bacteria, or insects. There are many different kinds of galls primarily caused by wasp, fly, and aphid insects, but although unsightly, they are not considered very damaging. Insects "sting" a plant, which causes a growth that the insect uses as a home for its young. The gall serves as a shelter and food supply. Natural control is biodiversity. Healthy plants seem to have fewer galls. Improve the general health of trees using the Basic Organic Program.

Gray Leaf Spot: A fungal disease of St. Augustine grass that forms gray vertical spots on the grass blades. A light baking soda or potassium bicarbonate spray is the best curative. Prevent by spraying compost tea and improving soil health. Treat the soil with cornmeal at 10–20 pounds per 1,000 square feet. Bio Wash (to clean plants) is also effective.

Herbicide Damage: Most, if not all, of the toxic chemical herbicides can cause plant problems. Contact killers and the preemergents can kill the beneficial fungi and the feeder roots of trees. Symptoms can look remarkably similar to oak wilt. That may be exactly what we are seeing in some oak wilt areas. Detox the soil with Garrett Juice and orange oil. To save trees and shrubs, use the Sick Tree Treatment.

Iron Deficiency; Iron Chlorosis: One of the nutrient deficiencies that causes leaves to turn yellow with the veins remaining green. At an advanced stage, leaves are completely yellow and then dead. Treat with greensand and compost. Insufficient iron in plants is characterized by striped, yellow, or colorless areas on young leaves. Yellow leaves appear on the new growth. The growth of new shoots is affected, and plant tissues may die. Too much lime causes iron deficiency to develop. Treatment: Plenty of manure, compost, blood or cottonseed meal, and greensand are the best materials to use in correcting iron deficiency. Magnesium deficiency looks identical. Most chlorotic plants have a lack of trace minerals in general.

Juniper Dieback: Fungal disease of cedars and junipers. Also called twig dieback. Spores look like a yellow worm oozing out. Treat by pick-pruning and mulching. Spray hydrogen peroxide and apply the Sick Tree Treatment.

Leaf Blisters and Curls: Fungal diseases that cause distorted, curled leaves on many trees. Oak leaf blister can defoliate and even kill oak trees. Blisters are yellow bumps on the upper surface of the leaves, with gray depressions on the lower surface. Peach leaf curl attacks peaches and almonds. New leaves are pale or reddish. The leaf's midrib doesn't grow, causing the leaves to become puckered and curled. Fruit is often damaged, and bad cases can kill the tree. Both diseases are controlled with a single horticultural spray just before buds begin to swell in late winter and another in autumn when the leaves start to fall.

Leaf Spots: Generic term for cosmetic diseases of oak and elm. No control is needed. A vast number of fungi can cause spots on the leaves of plants. Most of them are of little consequence. A typical spot has a definite edge and often a darker border. When lots of spots are present, they can grow together and become a blight or a blotch.

Lichen: Growth seen on rocks and the trunks of trees commonly in flat greenish, gray, brown, yellow, or black patches. Lichen consists of two separate and different microscopic plant forms—algae and fungi—that live together in a symbiotic relationship. The fungi absorb and conserve moisture and provide shelter, without which the algae cannot live. The algae conduct photosynthesis as they grow and provide protein for the fungi. Lichens are generally not harmful and mostly beneficial.

Lightning Damage: If your tree gets hit, keep your fingers crossed. Install lightning protection to prevent future damage. There are two kinds of lightning damage. When the lightning travels along the outside of the tree in the rainwater, bark is knocked off but damage is usually minimal. If the lightning goes through the center of the tree, the bark is blown off all around the trunk, and the tree is a goner.

Molds: Fungi that have a powdery or woolly appearance. *Botrytis* thrives in moist conditions and is often seen on dropped flower petals or overripe fruit. Look for a thick gray mold or water-soaked spots on petals, leaves, or stems. It first infects dead or dying tissue, so remove faded flowers and blighted buds or shoots to control the problem.

Peonies, tulips, and lilies can be particularly sensitive in wet weather. Remove and compost infected material. Space, prune, and support plants to encourage good air movement. Spray plants with Garrett Juice plus potassium bicarbonate.

Oak Leaf Blister: A rare disease that needs no control. It usually results from a heavy rainy season. Leaves change from light green to a light brown dry blister, and this only happens in isolated spots on the tree.

Oak Wilt: A disease of the vascular system of oak trees that is transmitted through the air by insects and through the root system of neighboring trees by natural grafting. Apply the Sick Tree Treatment.

Oak wilt attacks red oaks and live oaks, especially when they occur in large monocultures and are treated with synthetic fertilizers and pesticides. The disease on red oaks shows up first as greasy green leaves that then turn brown, starting on the ends—usually seen on one limb at a time. Live oak leaves have veinal necrosis (brown veins and green in between). Some leaves will be dead on the end half. Red oaks have sweet-smelling fungal mats on the trunks. Dutch elm disease is closely related. We do not recommend injecting fungicides into trees or removing trees that are near sick trees. The chemical injection hurts the tree, wastes money, and doesn't address the real problem. This procedure has been pushed by Texas A&M and the Texas Forest Service for many years, but we have yet to have anyone report that the fungicide injections have ever saved a single infected tree.

Biodiversity and soil health are the best deterrents. The disease is spread in the spring by a small beetle called nitidulid. Some people advise painting pruning cuts of live oak and red oak in the spring. We're not so sure this is important. If you do, use Lac Balsam or natural shellac.

Peach Leaf Curl: A fungal disease that causes leaves to be puckered and reddish at first, but later in the season turn pale green, shrivel, and drop. Fruit may have a reddish, irregular, rough surface and decreased production. This disease causes deformed leaves and can affect the quality and quantity of the fruit crop. Spray Garrett Juice plus garlic tea in fall.

Phytophthora: A genus of oomycetes that is generally a root and crown rot pathogen. This is a common disease in periwinkles that are being grown using synthetic fertilizers and pesticides. Beneficial microbes in healthy organic soil will normally keep this disease organism under control. Lilacs, rhododendrons, azaleas, and some hollies are sometimes infected by *Phytophthora* pathogens. Plants suffer shoot dieback and develop stem cankers. Prune to remove infected branches and to increase air movement. Add compost and rock powders to the soil. In healthy soil, plants rarely have this problem. On peppers, potatoes, and tomatoes, *Phytophthora* infection is known as late blight. The first symptom is water-soaked spots on the lower leaves that enlarge and are mirrored under the leaf with a white downy growth. Dark-colored blotches, sometimes like sunken lesions, penetrate the flesh of tubers. During a wet season, plants will rot and die. *Phytophthora* overwinter on underground parts and in plant debris. To control, dispose of all infected plants and tubers, presoak seed in a disinfecting solution such as hydrogen peroxide, and plant resistant cultivars. Sprays of potassium sulfate can help control outbreaks during wet weather. This is the disease that has greatly reduced the use of periwinkles as a bedding plant. Building healthy soil through the use of compost, natural organic fertilizers, and rock powders will control the pathogen. Good drainage is critical. Horticultural cornmeal is effective on this disease.

Pierce's Disease: A bacterial disease of grapes. It lives on but does not infect alfalfa, blackberries, cedar elm, grasses, red oaks, and willows. Leafhoppers are one of the main vectors. Management includes removal of host plants around vineyards and stimulation of beneficial microorganisms. This is a human-created stress problem caused by pushing the plants with pesticides and too much fertilizer.

Pine Tip Blight: A pine disease that forms black bumps on the needles. It resembles pine tip moth damage and can be controlled with the Sick Tree Treatment.

Potassium Deficiency: Potassium deficiency causes reduced vigor and poor plant growth. Frequently, the older leaves turn white and curl, later becoming bronzed. Severe deficiencies cause poorly developed root systems. To treat deficiencies, add rock, granite dust, wood ash, or other potassium-rich organic material to the compost and directly to the soil. Heavy consistent mulching also helps maintain the potassium supply.

Powdery Mildew: This fungal disease is a white or gray powdery growth on the lower leaf surface and flower buds of zinnias, crape myrtles, many vegetables, phlox, lilac, melons, cucumbers, and many other plants. It is common during cool, humid, cloudy days. Leaves turn yellow on the top. Controls include compost tea, baking soda spray, potassium bicarbonate spray, neem, garlic, and horticultural oil. Bio Wash (to clean plants) seems to be the best immediate cure. Garrett Juice plus garlic tea is best for long-term results. Treat soil with horticultural cornmeal and use the entire Sick Tree Treatment for serious problems. This disease can cause long-term weakness when it occurs early in the growing season. This common fungal disease is increased by humidity but actually deterred by water. Small black dots contain spores that are blown by wind to infect new plants. Leaves will become brown and shrivel, and fruits ripen prematurely and have poor texture and flavor. Prune plants to improve air circulation. Thoroughly compost infected plants before spores form and spread.

Pythium: A fungal pathogen that's a water mold, soil inhabitant, and common cause of root rot and damping-off of seedlings. Horticultural cornmeal is the best treatment.

Root and Stem Rot: These rots can usually be controlled with good drainage, good air circulation, and healthy soil. Only use healthy plants. Compost all infected plant material. Be careful to watch for physical injury, which may invite problems on woody plants.

Rust: Fungal disease that forms an orange stain on the surface of foliage. Pustules form on the underside of leaves. This is mostly a cosmetic disease but can be treated with Garrett Juice mixed with garlic or potassium bicarbonate. Rusts require two different plant species as hosts to complete their life cycle.

Scab: Fungal diseases that cause fruit, leaves, and tubers to develop areas of hardened and sometimes cracked tissue. Fruit scab can be a major problem on apples and peaches. It is recommended to dispose of fallen leaves, prune to increase air movement, and spray during the growing season with sulfur. Regular sprays of garlic and seaweed will also help. Compost tea is also beneficial, and horticultural cornmeal should be applied to the soil.

Slime Flux: Slime flux is the foul-smelling sap that oozes out of wounds in the bark and wood on trees such as elm, maple, and birch. The sap ferments and produces chemicals that kill the bark over which if flows. If the seepage continues over a period of a month or more, considerable injury and even death to the bark will occur. Control with the Sick Tree Treatment.

Slime Mold: Slime mold sounds worse than it is. Slime molds cover aboveground plants with a dusty gray, black, or dirty yellow mass. There will be tiny round balls scattered over the plant. If you rub these balls between your fingers, a minute sootlike powder will cover them. You can do nothing, or spray with garlic tea for control. Dusting dry cornmeal will also help. In turf, slime mold spore masses coat the grass and look like cigarette ash on the surface of the blades. The spores can easily be wiped off. Remove the mold spores from the grass by rinsing with water during dry weather or mowing and raking at any time. Baking soda spray will kill it, but potassium bicarbonate is better.

Smut: Fungal diseases common to grasses, grains, and corn. Corn smut attacks kernels, tassels, stalks, and leaves. Smut galls ripen, rupture, and release spores through the air to infect other plants and overwinter in the soil. Select resistant cultivars, remove and compost galls before they break open, and rotate crops. Spray Bio Wash (to clean plants) as preventative.

Sooty Mold: Black fungal growth on the foliage of plants such as gardenias, crape myrtles, and other plants infested with aphids, scale, or whiteflies. It is caused by the honeydew (excrement) of the insect pests. Best control is the release of beneficial insects to control the pest bugs. Also spray Garrett Juice plus garlic. Treat soil with horticultural cornmeal and use the entire Sick Tree Treatment. Bio Wash (to clean plants) is the curative treatment.

St. Augustine Decline (SAD): Disease in St. Augustine grass caused by the Panicum mosaic virus. It causes a yellow mottling, and the grass slowly dies away. The answer is to replace turf with a healthier grass. The best St. Augustine at the moment is "Raleigh." Switching to the organic program will eliminate the problem in most cases. Hydrogen peroxide at 3 percent has eliminated mosaic virus in squash plantings and may work on this grass disease.

Take-All Patch: A disease that can attack several species of grass. It is caused by the fungus *Gaeumannomyces graminis* var. *graminis* and is mostly found in St. Augustine grass but can also cause problems in Bermuda grass. It is most active during the fall, winter, and spring, especially during moist weather.

The first symptom is often yellow leaves and dark roots. Area of discolored and dying leaves will be circular to irregular in shape and up to 20 feet in diameter, and thinning occurs. Unlike brown patch, the leaves of take-all-infected plants do not easily separate from the plant when pulled. Stolons will often have discolored areas with brown to black roots.

Regrowth of the grass into the affected area is often slow and unsuccessful because the new growth becomes infected. Controlling take-all patch is said to be difficult, but it isn't with organic techniques. Good surface and subsurface drainage is important. Cut back on watering and fertilizing. Use only organic fertilizers and apply quality compost. If soil compaction exists, aeration will help to alleviate this condition and allow the grass to establish a deeper, more vigorous root system.

Prevent take-all patch by maintaining healthy soil. Control the active disease by aeration, cornmeal and compost applications, and the Basic Organic Program.

Tree Decline: A generic term referring to a sick, declining tree. Not a specific disease but rather a result of planting an ill-adapted tree, construction damage, drought, lightning strike, using salt fertilizer, chemical contamination, etc.

Viruses: The smallest and most difficult to control of all microorganism pathogens. Plastic mulch reflects ultraviolet light that repels insects that carry various devastating diseases, including viral diseases. University research has shown that using silver-colored plastic mulch under tomatoes and other plants can provide significant insect control. Although it may work, we don't really like it as much as natural mulch. Mosaic viruses destroy chlorophyll, causing leaf yellowing. Another virus blocks the plant's vascular system, restricting the flow of water and nutrients. Viruses are transmitted by vegetative propagation, in seeds, on pollen, and on tools and gardeners' hands. They are also transmitted by insects, mites, nematodes, and parasitic plants. Viruses slow plant growth and reduce yields. Infected leaves may deform and develop mottling, streaking, or ring-shaped spots. Identification is sometimes the elimination of all other possible causes. Purchase certified plants, control insects that spread viruses, and remove and thoroughly compost all infected plants. We have several reports of curing mosaic virus with a spray of 3 percent hydrogen peroxide.

Wet Wood: Bacterial wet wood on trees shows up as oozing cell sap, a white frosty material that attracts insects. Increase the tree's health so it can wall off the problem area. Use the Sick Tree Treatment.

Wilt: When fungi or bacteria clog a plant's water-conducting or vascular system, they can cause permanent wilting and death. Wilt symptoms may resemble those of blights.

Wind Damage: Don't overprune. Remember that heavy pruning is weakening and detrimental to tree health. Pruning is done in most cases for your benefit, not the tree's benefit. Research does exist now showing that proper pruning will help prevent trunk damage in severe wind events.

Wood Rot: These rots are usually the result of physical injury and normally don't kill healthy trees. Treat by removing decaying or dead wood and spray with 3 percent hydrogen peroxide. Don't cut back into healthy tissue; that only spreads the decay. Don't paint the area—leave it exposed to air to speed healing.

Xylella fastidiosa: Also called xylem-limited disease, it plugs the vascular system of elms, oaks, sycamores, pecans, and other trees. It looks like heat-stress damage and can easily be diagnosed by lab tests. Use the Sick Tree Treatment.

Weeds: The Disliked Plants

Finding anything positive to read about weeds is difficult. Unless you've read Malcolm Beck's *Lessons in Nature* or Charles Walters's *Weeds*, you probably haven't heard much good about weeds (see Information Resources for these and other publications).

Weeds are nature's greatest and most diverse group of plants. Even though many members of the weed fraternity are beautiful, people have been convinced by the chemical poison fraternity to condemn the weeds and consider them the enemy. Mention weeds, and most people think in terms of control through spraying toxic pesticides. They rarely think of why the weeds grow or of their value.

Weeds exist for very specific purposes. Different weeds have different jobs to do. Some are here to provide the soil a green blanket to shade and cool the ground. Others are here to prevent the erosion of bare soil. Others help balance the minerals in the soil. Many weeds provide all these important functions.

Weeds take no chances. They germinate and spread to protect any soil left bare from mismanagement of the land. In every cubic foot of soil lie millions of weed seeds waiting to germinate when needed. When the green growth is stripped off the land, weeds are needed. When hard winters freeze the ornamental lawn grasses, weeds are needed. When we mow too low and apply harsh chemicals to the soil, weeds are needed.

If it weren't for weeds, the topsoil of the earth would have eroded away years ago. Much of the topsoil has already gone from our farms forever to muddy our rivers and fill our lakes and eventually end up in our oceans. It's a common misunderstanding that weeds rob our crops of moisture, sunlight, and nutrients. Weeds only borrow water and nutrients and eventually return it all to the soil for future crop use.

Annual weeds grow best in soils with lots of bacteria and little fungi. Hence if one uses a broad-spectrum fungicide, this creates conditions favorable to the germination of weed seeds. Most perennial plants want lots of fungi, and some old-grown forests will require more fungi than bacteria.

Weeds are tough. Rarely do you find weeds destroyed by insects or disease. Some weeds are pioneer plants, as they are able to grow in soil unsuited for edible or domesticated plants. Weeds are able to build the soil with their strong and powerful roots that go deep, penetrating and loosening hard-packed soil. The deep roots bring minerals, especially trace elements, from the subsoil to the topsoil.

Weeds are indicators of certain soil deficiencies and actually collect or manufacture certain mineral elements that are lacking in the soil. This is nature's wonderful way of buffering and balancing the soil. Most plant species that we call weeds require their nitrogen in a nitrate form (NO_3), hence most synthetic fertilizers favor weed growth over desired plants. Weeds also require soils dominated by bacteria species, so when a fungicide is applied to the soil, the soil becomes bacteria dominated and favors the growth of weeds over desired perennials, shrubs, and trees that require lots of fungi in the soil.

Some weeds are good companion plants. Some have insect-repelling abilities, while others with deep roots help surface-feeding plants obtain moisture during dry spells. Weeds act as straws to bring water up from the deep, moist soil so that shallow-rooted plants can get some of the moisture.

Control becomes necessary when the vigorous weeds become too numerous in fields and gardens. However, not understanding the dangers of spraying toxic chemicals (like Roundup; 2,4-D; MSMA, etc.) into the environment, many farmers, gardeners, and landscapers have primarily used powerful toxic herbicides. Most herbicides upset the harmony of the soil organisms, and some herbicides can persist in the soil for months or longer. Even though microbes can repopulate after chemical treatment damage, they are slow to reestablish the complicated natural balance.

There are safe and nonpolluting weed-control methods such as mechanically aerating, mulching with organic materials, and using organic fertilizers to stimulate the growth of more desirable plants. The old, reliable methods of hand weeding, hoeing, and timely cultivating are not yet against the law and are good exercise. Goats contained by single-wire electric fences do an excellent job of brush control. Companies now exist that rent these animals. They are particularly effective for poison ivy eradication.

The best weed control in turf is the following: Water deeply but infrequently, use natural organic fertilizers, mow at a higher setting, and leave the clippings on the ground. Easy and effective weed control in ornamental and vegetable beds is done by keeping a thick blanket of mulch on the bare soil at all times. Clover, wild violets, and other herbs and wildflowers should sometimes be encouraged. Many plants that start out looking like noxious weeds end up presenting beautiful flower displays and wonderful fragrances. They're called wildflowers.

Weed control starts with a new attitude about weeds. A few are acceptable, even beneficial. On the other hand, some weeds need to be controlled. Here is a rundown of our suggestions for specific weeds.

Annual Bluegrass (*Poa annua*). Annual low-growing, cool-season weed. Prevent with corn gluten meal applied in the fall at 20 pounds per 1,000 square feet. Overspray in the winter with the vinegar or fatty-acid products.

Aster, Roadside (*Aster exilis*). Annual broad-leafed wildflower with white or light blue flowers in fall. Control by improving the moisture level and fertility of the soil.

Bermudagrass (*Cynodon dactylon*). Although this is a widely used turfgrass, we consider it to be one of the most troublesome weeds of all. You will, too, if it gets in your beds. Physical removal is the best approach. Never till it prior to attempted removal. Driving pieces of the rhizomes and stolons down into the soil will give you weed nightmares forever.

Bindweed (*Convolvulus arvensis*). Introduced from Eurasia, this plant is ranked among the dozen worst perennial weeds in the world. Roots go 6 inches deep and can lie in the soil for thirty years and still germinate. Also called wild morning glory. Control by increasing organic matter in the soil and using flame devices and vinegar sprays.

Black Medic. This legume, closely related to alfalfa, is an annual, biennial, or short-lived perennial. It is most often found in lawns having low fertility. Often called Japanese clover, this plant has small yellow flowers and a deep taproot. It can be pulled from moist soil without difficulty.

Brambles (*Rubus* spp.). Various berry plants with sharp thorns that spread to form dense masses. Control by pulling up. Spray regrowth with vinegar-based or fatty-acid herbicide.

Bull Nettle (*Cnidoscolus texanus*). Perennial problem weed in deep, sandy soils with low fertility. Leaves and stems are covered with stinging hairs. Huge underground storage tubers. Control by increasing organic matter in the soil.

Bur Clover (*Medicago hispida*). Very low-growing annual cool-weather legume. Small yellow pealike flowers. Seeds are contained in a soft-spined bur. Control by increasing soil health. It can be killed by spot-spraying the vinegar formula or a fatty-acid product.

Canada Thistle (*Circium arvense*). Perennial weed, 1½–4 inches in height. Very difficult to control because of its deep root system. Control by mowing when plant is in full bloom. Root system is exhausted when it is the prettiest.

Chickweed (*Stellaria media paronychia*). Annual cool-season broadleaf weed with small, bright green leaves. Low-growing benign weed. Just mow it. Spray with vinegar or fatty-acid product.

Cocklebur (*Xanthium strumarium*). Tall, bushy annual weed with prickly seeds and sandpaper-like leaves. Grows where excess phosphorus is available. Compost and organic soil amendments are the main solution.

Dallisgrass (*Paspalum dilatatum*). Long-lived warm-season deep-rooted perennial bunch grass. Forms low, flat clumps with dead-looking centers and tall fast-growing seed heads. Chop out mechanically and fill holes with compost. MSMA, an arsenic compound recommended by those not concerned with toxic herbicides, is in the process of being taken off the market.

Dodder (*Cuscuta* and *Grammica*). Annual weed that reproduces by seed. It starts as an independent plant but establishes a parasitic relationship with the host crop. At this point it has no chlorophyll and looks like yellow string. A twining yellow or orange plant sometimes tinged with purple or red; occasionally it is almost white. The stems can be very thin and threadlike or relatively stout. Dodder rarely kills its host plant, although it will stunt its growth. Control by balancing the minerals in the soil. It can be killed with selective sprays of vinegar herbicides.

Field Bindweed. See *Bindweed.*

Goathead (Puncture Vine; *Tribulus terrestris*). Hairy, low-growing annual with a taproot and several stems forming a rosette. Has yellow flowers and burs that will puncture tires. Same control as grassburs. Build soil health. Apply corn gluten meal or mustard meal.

Goosegrass (Silver Crabgrass; *Eleusine indica*). Annual that reproduces by seed in unhealthy soil. Very similar to crabgrass. Build soil health.

Grassbur. See *Sandbur.*

Henbit. A species of *Lamium* that many consider a wildflower. Control, if you must, by mowing or spot-spraying with vinegar-based herbicides.

Honeysuckle (*Lonicera japonica* 'Atropurpurea'). Physically remove this invasive vine.

Mistletoe (*Phoradendron flavescens*). Plant parasite that primarily attaches to limbs and trunks of low-quality and/or stressed trees, such as Arizona ash, hackberry, bois d'arc, locust, boxelder, and weak elms and ashes. Remove by cutting infected limbs off the tree. If that can't be done, notch into the limb to remove the rooting structure of the mistletoe and paint with black pruning paint to prevent resprout. There are no magic chemical or organic sprays. Keeping the soil and trees healthy is the best preventative.

Nutgrass (*Cyperus rotundus*). Perennial sedge introduced from Eurasia. Spreads by anaerobic seed, nutlets, and creeping tendrils. Likes wet soil. Remove with mechanical devices. Control in turf by planting ryegrass in the fall. Heavy applications of molasses help. Use ½ to 1 cup of liquid molasses per 50 square feet. Solving drainage problems and eliminating an aerobic soil is very important.

Oxalis (Creeping Woodsorrel; *Oxalis corniculata*). Leaves of oxalis have a shamrock appearance, and the plant is often mistaken for a clover. At night or on cloudy days, the leaves may fold up. Leaves turn purplish in color in cold weather. Some plants may have purple leaves year-round.

Prostrate, creeping perennial weed with stems that will take root where they touch the ground. Flowers are small and yellow. Mature fruits explode, scattering seed several feet away. Can be killed by spot-spraying with vinegar or fatty-acid products.

Pigweed (Redroot Pigweed; *Amaranthus retroflexus*). Redroot pigweed is an annual weed commonly found in waste areas and disturbed soils. Grows 2–4 feet high. Lower stems are reddish in color, and flowers are small and green. A prolific seed producer, pigweed will produce up to 100,000 seeds per plant. Seedlings are easily pulled or hoed from the garden. Mulch will prevent germination of seeds already in the soil. Spray growing plants with vinegar or fatty-acid products.

Poison Ivy (*Rhus radicans*). Deciduous vine that grows in sun or shade and spreads easily underground. Has red berries and red fall color. Do not allow to flower and produce seed. Remove, compost, and spray new growth with vinegar-based organic herbicides.

Purple Nutsedge. See *Nutgrass.*

Ragweed (*Ambrosia* spp.). Annual broad-leaved plant that indicates droughty soil. Releases a potent pollen that causes hay fever. Control by cultivation, mowing, building the soil, and spraying with vinegar-based organic herbicides.

Rescuegrass (*Bromus cathearticus*). Cool-season annual bromegrass. Control by broadcasting corn gluten meal in early October or before seed germinates.

Sandbur (*Cenchrus pauciflorus*). Annual grass plant that produces a bur with strong, sharp spines. Seeds in the bur can lie dormant in the soil for years before germinating. Control by increasing the carbon in the soil with compost, humates, dry molasses, and/or corn gluten meal.

Smilax (*Greenbriar*). High-climbing perennial that grows in sun or shade. Flowers from February to June and produces black berries in September and October. Woody vine with strong, thorny stems from large underground tubers. Leaves are tardily deciduous and sometimes white-blotched. Deer like it. The tender new growth is delicious in salads. Control by digging out the woody underground tubers.

Spurge (*Euphorbia* spp.). Sappy, succulent annuals or perennials that like hot, dry weather. Control by spot-spraying vinegar-based organic herbicides.

Preemergent control of all weeds growing from seed is done by the application of corn gluten meal at 20 pounds per 1,000 square feet before the germination of the targeted seed. In general, the timing is February 15–March 15 in the spring for summer weeds and September 15–October 15 in the fall for winter weeds. It can also be used after tilling or otherwise worked into soil to prevent the growth of seed in disturbed soil. Corn gluten meal has fallen out of favor because of the high cost of corn and the inconsistent efficacy of its herbicidal power.

Organic Herbicides

Two versions of vinegar-based herbicides can be found in Appendix 1. Let us know which one works best for you.

Vinegar strengths vary: 5 percent is regular vinegar, and doesn't work well for weed control; 10 percent is pickling vinegar; 20 percent is too strong and no longer recommended.

There are acceptable fatty-acid products as well. Two commercial products are Scythe and Monterey Herbicidal Soap.

CHAPTER 6

Compost

The purpose of this chapter is to introduce readers to compost and give them an understanding of how it is made. There are huge differences in the performance and value of the many products sold as compost. Additionally, many states do not have labeling laws or quality standards, so many worthless products are sold under the label compost. We summarize the benefits of good compost, separate the products into three grades for reference, give a little history of the composting industry, and then explain to the reader how to tell good compost from bad. The last section covers variations of compost (e.g., vermicompost) and "compost tea" and gives readers an introduction to both.

Why Compost?

- Compost provides primary nutrients as well as trace minerals, humus, and humic acids in a proportion that almost exactly matches plant requirements. Compost also helps unlock minerals present in existing soil. Why waste money on expensive artificial fertilizers when you do not have to?
- Compost is a food source for earthworms, beneficial insects, and microorganisms that improve soil structure and prevent disease. For example, compost contains nematode-destroying and other beneficial fungi used as a treatment for many plant diseases such as brown patch and damping-off. Why spend money on dangerous synthetic chemicals?
- Compost helps tight clay soil become more friable and breathe and helps loose sandy soil hold more water and nutrients. Compost helps increase air spaces and makes the soil more resistant to compaction. Why spend time and expense for extra watering and core aeration?

- Compost contains growth-promoting hormones for plants that encourage the development of an extensive healthy root system (wider and deeper), which helps plants tolerate drought conditions, insects, and disease. Why spend money on dangerous chemicals and replacing valuable plants?
- Compost acts as a buffer against chemicals and absorption of dangerous heavy metals by plants. Microbes in compost break down hazardous chemical pollutants—from pesticides and herbicides to oil and diesel fuel. Using compost makes your yard safer for children to play in and for family pets. Why worry about dangerous chemicals in your managed landscape?
- Compost helps buffer soils against extremes in acidity or alkalinity (high or low pH), and properly prepared compost tea is an excellent foliar fertilizer and natural fungicide. Why should you have to frequently add other expensive inputs to your soil?
- Each year we use compost, the soil becomes healthier and we get better plants. Using quality compost saves consumers lots of money, with the benefits increasing each year. Since compost is made from 100 percent recycled materials diverted from landfills or incineration, using compost benefits the environment. Why haul organic material to the landfill?

Compost Types and Methods

There are many ways to produce and make compost. Two main methods of commercial composting use windrows or static piles. The windrow method creates smaller piles and uses a machine (sort of like a giant garden tiller) to run through the pile every few days or more to aerate it. The smaller pile size and forced aeration does not allow the anaerobic microbes to grow, and the constant turning of the pile rips the required fungus hyphae to shreds, killing the fungus, hence we lose the two major components required to break down herbicides. The windrow method is very fast, producing compost in only 4–8 weeks, thus there is not enough time for the remaining microbes to even begin to break down any herbicide that might have been present in the original feedstock material.

The other method of commercial composting is called the static pile. Feedstocks are mixed and placed in large piles from 10–12 feet tall to over 50 feet tall, 25–30 feet wide, and hundreds of feet long. These piles are only turned a few times from start to finish, hence the required good fungus is allowed to grow and colonize the piles. Small areas will form inside the big piles that will become anaerobic. Some of the microbes (facultative microbes) will change their metabolism to work in the low-oxygen environment. This change is what allows them to break down the herbicides and other material, after which oxygen can reenter the pocket and the microbes switch their metabolism back to using the oxygen.

Many states do not have labeling laws or compost standards, which allows many products to be sold as compost. These products can be broken into three basic types, based on their quality and usage: biological, commercial, and industrial.

Biological: This is the highest-quality compost and therefore the most beneficial for improving soils, preventing disease, and making compost tea. This is the product that experienced gardeners often call *black gold* because it is so valuable to plants. Retail bags will have holes so that air can enter and the beneficial microbes can breath and be kept alive.

Commercial: This is a middle grade of compost made from sewage sludge, construction debris, manure, or other materials. It will be in a sealed bag and may have a sour or stale odor. The better manure-based composts may be found here.

Industrial: This is the lowest grade of products called compost (not a real compost). It is made from industrial wastes like boiler ash and is often very black and sometimes will rub off in your hand. It often contains fillers like sawdust and rice hulls, which are chemically burned black from the industrial waste. It may be extremely alkaline and high in toxic salts.

Note: The price of good compost varies greatly around the country. The cost of producing compost is driven by local landfill disposal charges (dump or tipping fees) and local environmental regulations. The U.S. Composting Council (USCC) regularly produces a compost price survey in their monthly newsletter, based on feedstock used and by geographical area of the country. It does not cover quality or processing but does give an idea of regional costs. These will be explained in detail in later sections.

What Is Compost or Composting?

Composting is, in the simplest terms, the biological reduction of organic wastes into humus. It is a natural process used to return nutrients contained in organic materials that were once alive to the soil. Compost is composed of complex carbohydrates, proteins, amino acids, peptides, humic acids, fulmic acids, enzymes, and beneficial microorganisms. Whether composting takes place in a compost pile or on a forest floor, it is the same clean, sweet, earthy-smelling material that is so essential for healthy plant growth.

Is Composting a New Process?

Nature was making compost eons before humans ever walked on earth. References to using manure and wheat straw in composting were common in biblical times. In fact, the Greek and Hebrew words for compost piles were originally translated "dung piles." Recent archaeological discoveries clarify that the dung piles referred to in the Bible were indeed compost piles. References to composting have been found on clay tablets dated thousands of years ago.

Compost was used by the Greeks and Romans to boost crop production, and they even used the heat of decomposing materials to produce summer vegetables in winter. Arab writings of the tenth and twelfth centuries include passages on how to compost and its importance to agriculture. A Christian monastery kept the art of composting alive in Europe, and by about A.D. 1200, compost was again being used by many farmers. Shakespeare mentions it in several of his plays written in the early 1600s. The word *compost* comes from the Medieval Latin *compostum*, which means "bring together."

The first recorded commercial composting operation for municipal refuse was started in Holland in 1932 and is still in operation today. The EPA (U.S. Environmental Protection Agency) estimates that there are over 3,000 composting operations in the world outside of the United States. The EPA also estimates that there are over 10,000 composting operations of all sizes in the United States.

Compost is produced by the breakdown and conversion of materials such as grass and tree trimmings, food waste, scrap paper, cardboard, manures, scrap wood, liquids, biosolids, and others that are composed of organic compounds from materials that were once alive. Usually these materials are passed through a grinder, with moisture and microorganisms added, and then placed in some form of pile to undergo an accelerated and controlled decomposition process.

The decomposition takes place by the microorganisms (bacteria, fungi, etc.) literally eating this material and converting it back into simpler and different chemical compounds. These microorganisms take oxygen (and some nitrogen) from the air and combine it with carbon from the raw material and water to produce compost, CO_2, steam, and heat. The exact decomposition involves the chemical and physical processes of oxidation, reduction, hydrolysis, and thermodynamics.

Compost can vary widely in nutrient content, disease-fighting ability, humus content, and other characteristics. These factors are determined by the type of source materials, the mixture (ratios) of this source material (chemistry), the temperature and structure of the piles (physics), the time the pile stays at certain temperatures (physics), and other things. A knowledgeable practitioner can control the composting process to produce customized compost. For example, one might increase the nitrogen content for use as fertilizer or maximize the content of beneficial microorganisms for use as disease control in greenhouse or nursery potting soils. As for any other product, compost quality ranges from excellent to very poor, depending on the goals and knowledge of the operator.

The process of composting is approached from basically two points of view. The first is waste volume reduction to avoid using up valuable landfill space. The second is a manufacturing process to produce a valuable end product for use in agriculture, landscaping, and horticulture. The purpose and methods chosen for composting greatly affect product quality, cost, and odors produced.

The waste management industry generally implements the first point of view, in which the goal is to reduce the volume of waste. Engineering techniques are chosen (windrows, in-vessel, etc.) that cause the material's volume to decrease. This is done by converting the solid particles in the waste to gases (water vapor, odors, etc.) and releasing them into the atmosphere. In general, these types of processes offer fast volume reductions (less than twelve weeks from start to finish) and require very expensive special equipment. The resulting product often has fewer nutrients, since many useful chemicals have been released into the air to achieve the volume reduction.

The second point of view emphasizes composting as a manufacturing process in which the goal is a useful and valuable end product. Simple engineering is generally used, such as creating large static piles that are designed to prevent the loss of volume. Gases produced are absorbed (eaten) by other microorganisms and converted back to useful solids (nutrients). Longer time frames (six to thirty-six months) are required, as these piles are left undisturbed for weeks at a time (some operators go for one year or more) before turning. There is less volume reduction and less gas produced, hence less potential for odors.

To understand the differences in quality and value in the types of compost being sold, one needs to understand the basics of their manufacture or production. There are two general methods to make compost, aerobic (with oxygen) and anaerobic (without oxygen). The anaerobic methods are more common in Europe, with only a few operations in the United States, hence they are not covered in detail, as there are many textbooks and papers on the subject (see Information Resources). One form of this method is called Bokashi, developed in Japan.

The most common method in the United States is the aerobic technique in which raw organic materials are placed in piles to decompose. Within this category there are several variations that affect the quality and the cost of compost. These use oxygen and heat (thermal techniques) to decompose the raw organic materials.

The first method is to mix and place the organic materials (feedstock) into long piles called windrows. These are typically 8–10 feet wide at the base and 4–6 feet tall and may be of any length. These rows are turned by front-end loaders or special machines called windrow turners every few days or more often. The feedstock gets very hot and quickly decomposes due to bacterial action. Many of the nutrients in the material are converted to gases such as carbon dioxide, ammonium, and hydrogen sulfide, thus releasing nitrogen and sulfur into the air as well as producing odor. The frequent turning kills many of the beneficial fungal species that prevent diseases in plants and turfgrass.

The second method is known as in-vessel, in which the feedstock is placed into a rotating or auger system with air pumped into it. These operate with a continuous flow of material from start to finish. Again, however, the constant turning and mixing kills the beneficial fungi. Compost can be produced in two weeks or less with many of the issues associated with windrows.

The third method is known as the static pile, of which there are several variations. Feedstock is placed in large piles measuring from 8–15 feet tall and 15–25 feet wide at the base up to 50–60 feet tall and 100 feet wide (giant static piles). These piles can be of any length.

In some operations, the piles are placed over perforated pipe and air is pumped into the pipes to flow into the pile (forced aeration) to speed the decomposition. The piles are then turned every few weeks until the compost is ready in a couple of months.

The highest-quality compost comes from static piles that are allowed to decompose slowly from many months to several years. These piles are only turned a few times over the composting process. Aeration is done passively. Heat and moisture are released from the top of the piles' surface (hot air rises), and fresh air is pulled into the piles from the bottom and sides. The long time frames and occasional turning allow the beneficial fungi that prevent diseases to grow. The heat kills any pathogens that might have been present and any seeds in the feedstock. Another benefit is that any insecticides and herbicides that were in the feedstock will be broken down by the microbes (primarily fungi).

Often grass, leaves, animal manure, and bedding, etc., are used as feedstocks for composting, and these may have been sprayed with toxic chemicals, including herbicides and pesticides. There is no exact easy answer as to whether these chemicals will be broken down into harmless components without knowing which pesticides or herbicides, and in what quantities, are present. According to the University of Illinois Center for Solid Waste Management and Research, some common herbicides can stay active for one full year. Scientific research on how pesticides break down in the compost pile is just beginning, but the consensus is that some of these products might survive the composting process if short time frames are used (many composts have been found to have phytotoxic effects). Whether they do or not depends on which chemical was used, as different chemicals break down at different rates. Conditions within the compost pile (heat, moisture, pH, etc.) also affect the rate at which toxins will disappear, hence the composting method used will also directly affect the breakdown rate. For example, herbicide and insecticide molecules are more readily degraded under combined aerobic and anaerobic conditions (found in large static pile techniques), and the longer the time frame used (large static pile techniques) the greater the breakdown. Many of the potentially difficult chemicals have a half-life in a compost pile of only three months, hence if longer time frames are used, these chemicals are broken down below measurable limits.

Turning is the term used to describe the mixing process during the manufacture of compost. It adds air to the piles, mixes outside material deep into the pile, and loosens up the pile to help aeration. This may be done by front-end bucket loaders, windrow turners that straddle the row, or other specialized equipment.

Screening is the process of putting the finished compost through a machine that separates the compost into different sizes. They work like a giant flour sifter, only passing the desired size of material. This process removes larger pieces that may not have been fully composted. The most common screens are called trommel screens. Flat-top vibratory screens and other types are also used.

When compost is ready to use it is called *mature*. There are several methods used to measure maturity. The most common is the Solvita Compost Maturity Test developed by the Woods End Research Laboratory. A small amount of compost is placed into a plastic container. An indicator on a plastic stick is then placed into the container and the container is sealed. After a period of time, the indicator will change color. One then compares the color on the indicator to a numbered chart for each color. The degree of maturity can then be read from the indicator. Additional information can be found on their Web site at www.woodsend.org.

The most accurate method is to send a compost sample to a biological laboratory for testing, which takes seven to ten days to get your results. The lab can measure maturity and the amount of microbes in the product (bacteria, fungi, protozoa, nematodes, etc.), which indicates the value of the compost. More information can be found at www.soilfoodweb.com. *Note:* A few of the certified biological labs can also do chemical analysis at the same time.

Compostable Materials

The final properties of compost are determined by the materials being composted. Anything that was once alive may be composted. The following is a partial list of compostable materials from various industries and sources (C = carbon source, N = nitrogen source).

For a list of compostable materials, see: www.natureswayresources.com/fdaccepted22.html.

Livestock/Animals

- manure and bedding from zoo, rodeo, racetrack (C, N)
- horse manure and bedding (wood shavings) from stables (C, N)
- sewage sludge and sludge cake from wastewater treatment (N)
- chicken/turkey litter (N)
- cow manure and bedding (C, N)
- molasses and sludge residues from feeding tanks (N)
- animal mortalities: dead animals such as chickens and other poultry, cows, fish, city and county animal control, etc. (N)

Agricultural

- rice hulls (C)
- cotton burr wastes (C)
- pecan hulls (shelling operations), pruning from pecan orchards, and other wastes (C)
- peanut shells, coconut shells, etc. (C)
- eggshells, shrimp shells, crab shells, etc. (N)
- oysters and shells (minerals)
- grain dust (N)
- bagasse (sugarcane waste) (C)

Food Processing

- wastes from companies, packing houses, distribution centers, etc.
- coffee grounds, teas, chocolate, etc. (N)
- dairy by-products (cheese, whey, etc.) (N)
- bakery wastes (bread, cakes, crackers, etc.) (N)
- diatomaceous earth from filters
- beverages, alcoholic and nonalcoholic, including out-of-date beer/alcohol from distributors and confiscated alcohol
- fruits and vegetables (N)
- grain and grain dust from mills, elevators, haulers (trucks & barges), etc. (N)
- sugar and confectionery wastes
- meat and poultry (N)
- seafood and fish offal (N)
- slaughterhouse wastes and renderings: hooves, hides, horns, hair (N)
- food-processing waste, sludge, and wastewater (N)
- off-spec food and out-of-date food (N)
- nutshells, hair, feathers, seeds, etc. (N)
- bycatch from fishing industry (N)
- paunch manure (N)

Grocery Stores

- produce (N)
- deli products (N)
- bakery products (N)
- nonrecycled cardboard (wet or wax coated) (C)
- meat, bones, scraps (renderings) (N)
- out-of-date food, beer, juice, etc. (N)
- grease-trap waste (requires permit in some states) (N)

Industrial/Government

- wood pulp from sludge and/or wastewater (C)
- sugar and syrup wastes (N)
- paper from document-shredding companies (C)
- grease-trap waste (requires permit in Texas and other states) (N)
- wax-coated cardboard (C)
- wood, short-fiber pulp from paper companies (C)

- granite dust from stone-carving companies (minerals)
- drywall (sheetrock) from new construction waste (C, Ca, S)
- seaweed from beach cleanups (N, minerals)
- brewery waste (N)
- winery waste (N)
- distillery waste (N)
- pallets and shipping containers (C)
- sawdust and construction wood scraps (homes, cabinet shops, lumberyards, etc.) (C)
- fabric/garment scraps (wool, cotton, etc.)
- cotton fluff from diaper manufacturing
- ATF-confiscated alcohol
- out-of-date beer, wine, liquor, etc.
- cotton lint from cleaning/manufacturing
- telephone books (C)
- junk mail from post office (C)
- pharmaceutical residuals (some have high nitrogen and protein content) (N)
- grain dust (barges, elevators, trains, trucks) (N)
- floral waste and trimmings (N)
- sewage sludge (biosolids) (N)

Sewage Sludge (Biosolids)

Biosolids are the digested and processed solids (mainly dead microbes) from sewage treatment facilities. Most states and the EPA have very strict guidelines on the composting of biosolids. This feedstock is nutrient rich; however, it can easily be contaminated with heavy metals, antibiotics, hormones, and literally hundreds of chemicals that get flushed down our sewers. Some of these are destroyed by the composting process, but many are not.

Composting of sewage sludge can be done, but the factors involved are much more complex both from a scientific point of view and with regard to governmental regulations. The potential for foul odors is higher, the cost is higher (extra government regulation), and the chance of pollutants in the sludge is higher. However, with proper planning and site preparation, the extra problems can be overcome. Several studies have shown that compost made with sewage sludge can be of good quality and beneficial, *if* properly done. The risk in using sewage sludge is that viruses can survive the high temperatures for some time, and complex chemicals such as polybrominated biphenyls (PBBs), heavy metals (lead, cadmium, arsenic, etc.), and pharmaceuticals cannot be easily removed from the compost. If the material entering the sewage system is regulated at the source, preventing contamination from occurring, then composting can be a very good solution and composting over a long time frame ensures that these chemicals are broken down. Most modern water treatment facilities, in compliance with current regulations, produce a sewage sludge ideal for composting. Many communities have found co-composting of sludge with ground brush or leaves to be an excellent solution (preferred over land application or landfilling), transforming a nasty waste disposal problem into a beneficial product. A good example of this is the city-run biosolids composting program of Austin, Texas, whose end product is marketed under the trade name Dillo Dirt. Dillo Dirt starts its process in windrows to meet governmental requirements but then the compost is allowed to mature (cure) in static piles to allow beneficial microbes such as good fungi to grow.

Biosolids compost is a commercial-grade product that does have beneficial uses if applied appropriately. It is appropriate for revegetation projects, erosion control, highways, sides of retention ponds, mine waste reclamation, bioremediation, turfgrass farms, and as an ingredient in nursery potting media and in many other industrial and commercial projects where only one application is required. It should never be used for any application where food crops are grown or in a 100 percent organic program.

Municipal solid waste (MSW) is another feedstock that is sometimes used for composting. The garbage is sorted, removing most of the plastics, glass, metals, and other noncompostable items. The remaining portion of the garbage is then composted. This method was popular in Europe many years ago but has fallen out of favor due to all the problems it created.

Compost as a Microbial Inoculant

The best inoculant products on the market have only two hundred species of bacteria and twenty species of fungi, versus good compost that will have twenty-five thousand species of bacteria, ten thousand species of fungi, as well as many species of protozoa and beneficial nematodes. These organisms are important as a soil inoculant that cycles nutrients and prevents disease. Mycorrhizal fungi, however, are not present in compost, as the heat of the composting process eliminates most of the mycorrhizal fungal propagules. Quality compost can have a great value as an inoculant. This is why good compost at a rate as little as 1 ton per acre can make a large difference for nutrient cycling and biological control.

Salt and Disease Reduction

Microbes in compost reduce the negative effects of high sodium (salt) in soils, reducing compaction and hardpan. The bacteria use sodium (Na) in their cell walls, hence removing it from the soil profile as an active chemical, and the fungi use chlorine (Cl) in their cell walls, thus removing salt from the soil profile as an active chemical.

The good microbes found in compost suppress many plant and soil diseases by competing for nutrients, producing chemicals that inhibit the growth of pathogens, or consuming (eating) the disease-producing microbes. Best results occur when the compost has high levels of fungal biomass and species diversity.

The microbes improve nutrient retention in soil and also produce enzymes that help mineralize nutrients and make them available to plants. Microbes improve soil structure both physically and chemically (more water and oxygen). They break down (decomposition) toxic materials (phenols, tannins, pesticides, herbicides, etc.), creating healthy, fertile soil. Many microbes also produce plant growth-promoting compounds and improve crop quality (flavor, nutrients, and yield).

Note: It may take 2–3 cubic yards more of a lesser-quality compost to give the same numbers of beneficial microbes as in good compost, hence it is more cost-effective to purchase the highest-quality compost one can find.

Composting Process

The making of compost has been described as both a science and an art. The truth is that it is a combination of both. Dead things are going to rot and become compost. Managing that process can affect the time and the quality of the finished product.

Processing Location

Compost can be made in full sun, full shade, on soil, on concrete, or on any other surface. Putting once-living materials into a pile is all that needs to be remembered. Nature will turn those materials into compost. It is not necessary to put compost piles in the sun or to cover them with tarps to help build up the heat in the piles. The highest microbial activity is inside the pile, not out on the outside edge. Covers may block oxygen exchange, which is critical for good compost.

Properly made compost with beneficial fungi visible. Photograph by Howard Garrett.

Containers

The most efficient manner of composting is not to use a container at all because it limits the volume. One of the problems people have in using containers is that they don't realize how much volume of raw material is needed to end up with an acceptable amount of finished product. Raw composting materials will shrink as much as 70 percent from the beginning to the end of the composting process. Unless space is a problem, containers are not recommended. Containers can also restrict air movement. If they are made of wood, they will compost just like the pile ingredients.

Turning of Piles: The process of turning or mixing compost piles is beneficial *if* not done too many times. Turning kills the beneficial fungi required for disease control. As a result, piles become dominated by thermophilic bacteria, causing the piles to break down quickly and become very hot (often over 170°F). Many beneficial microbes die off, and nitrogen and sulfur are released as gases, lowering the value of the compost. A temperature of 130°F during processing is needed to kill root-feeding nematodes; however, most mycorrhizal fungal propagules are also killed at this temperature. Good bacteria and fungal-feeding nematodes go dormant at 130°F and do not die until 170°F is reached. Similar temperature thresholds exist for some beneficial fungi and protozoa, too. The heat tolerance of microbes varies around the country, as microbes from southern climates (with much hotter air and soil temperatures) tend to be far more heat tolerant than microbes from northern climates.

Time Frame: many toxic chemicals that may be present in composting feedstocks take so much time to break down that it is measured as a half-life (the time needed for half the amount of a substance to disintegrate or be eliminated). Longer time frames allow for slower composting, ensuring greater diversity and abundance of microbes and nutrients while allowing complete removal of any possible contaminants. Herbicides are used on many plants—from hay and straw for animal feed to turfgrass in lawns. This is a major problem for some producers and users of compost. Many new herbicides are very resistant to breaking down and may cause problems for users of low-quality compost. Clopyralid and picloram are commonly used on fields to grow hay for cattle feed. A cow can eat the hay and the cow's urine will kill plants. The same problem occurs in the cow manure. The half-life of these herbicides in a compost pile is about 90 days *if* fungus is present. Hence, if fast composting is used, these chemicals are often still present and may cause problems when the compost is used.

Another herbicide that might survive and be found in quickly made compost is atrazine, a triazine herbicide. According to the National Toxicology Program (NTP), it is "immunotoxic," meaning it disrupts the function of the immune system (for example, it decreases interferon production, which fights viral infections). According to the EPA, testosterone, prolactin, progesterone, lutenizing hormone, estrogen, and thyroid hormone are all affected. According to the University of Iowa, babies may have low birth weights and a higher risk of birth defects when exposed to water with atrazine levels well above the EPA limits. In laboratory studies, it also caused genetic damage in animals. Other studies have shown a link between atrazine exposure and certain types of cancer.

A new herbicide, aminopyralid, has recently entered the market. It is so toxic that as little as 10 ppb (parts per billion) can harm plants, according to researchers at Ohio State University. This herbicide has a 533-day half-life in the soil. As a result, thousands of gardens have been harmed from compost made from feedstocks (manure, hay, etc.) treated with this herbicide. (See http://www.natureswayresources.com/resource/infosheets/killercompost.html for more information.)

Feedstocks: The materials used to make compost will affect the quality and determine the type of usage and amounts required to achieve a given result.

Arsenic is sometimes fed to poultry to make them gain weight faster. This toxic metal tends to accumulate in the manure. Hence some poultry manure composts may not be ideal for some uses.

Cattle manure from feedlots can contain pesticides, hormones, and drugs like ivermectin, which is given to cows (allegedly) to protect them from internal parasites. These drugs are toxic enough to kill insects that eat and process dung, preventing its decomposition, and then it kills plants. Thorough composting is very important to neutralize these possible contaminants.

Manure from horses, sheep, zebras, camels, etc., is a better material to use, as these animals have a much wider diversity and larger number of beneficial microbes in their guts, hence they tend to produce higher-quality compost.

Green waste consists of grass and leaves from a municipal recycling program. Generally this makes a superior feedstock, since it comes from yards that are well fertilized and watered.

Good compost can be made from any feedstock; it just depends on the objectives of the producer and the quality of the composting operation. Objectives range from getting rid of a waste product as quickly as possible (feedlot or slaughterhouse) to removing organic material from a landfill to save space (waste company) to manufacturing products for use in horticulture. Different objectives require different methods to produce and store compost. Unfortunately, a lot of vendors are selling products that are not compost or are compost that has been stored and handled incorrectly. You may be able to avoid getting a poor-quality compost by being aware of some of the warning signs.

Warning Signs in Compost

Odors: Bad compost often contains anaerobic organic acids that have a strong odor from putrefying organic matter. The odor varies depending on the feedstocks and what is going on, but they are all very bad.

Acetic acid produces a vinegar smell, and when present, it indicates the loss of nitrogen and phosphorus and the presence of alcohols from the anaerobic conditions (i.e., fermentation). Butyric acid has a sour milk smell and indicates alcohol is present. Valeric acid has a vomit smell and also indicates the presence of alcohol. Putrescine, another organic acid, smells like rotting meat and means alcohol is present. Alcohol is very damaging, as only 1 ppm of alcohol kills the roots of most plants.

Ammonia is another common odor, and it implies an immature compost (phytotoxic) and a loss of nitrogen in the ammonia. A rotten egg (H_2S) odor implies immature compost (phytotoxic) and a loss of sulfur.

Color: Pure "black" compost does not have good fertility and indicates anaerobic decomposition (fermentation) or other problems. A deep black color only occurs in nature when anaerobic conditions are present. Sulfur is gone (H_2S), nitrogen is gone or in the wrong form, and alcohols are usually present. Good compost is a deep chocolate brown when dry.

Industrial wastes are often used to blacken products for marketing purposes. (For example, even though it is illegal in many states, some companies grind up old railroad ties to help darken the material.) Others use dyes or pigments to blacken material and then call it compost for marketing purposes (see "Colored Mulch" in Chapter 7 for more details). Smelter wastes are sometimes used as a feedstock to blacken products. Copper sulfate ($CuSO_4$) or other sulfur compounds may be present. Elemental sulfur (S) is a natural fungicide and kills the beneficial fungi. Boiler ash (bottom ash) is another industrial waste product used to color or blacken products. Boiler ash is high in salts and extremely alkaline. The alkalinity is so strong that it will chemically burn raw wood black in a couple of days. Products tend to be alkaline when they have a high salt content, with very high carbon to nitrogen ratios. Some ashes may contain large amounts of heavy metals. Products will often turn a bleached grayish color in a few weeks. These types of products are very common in some areas of the country.

Fillers: Some producers cut their product with pine bark (a worthless, nutritionless filler often burned black by boiler ash) and may include sand or both. Rice hulls and rice hull ash are common fillers used to cut compost. Others grind up old wooden pallets, dye the material black, and then use this as filler and call it compost.

Tricky Names: Many products that are called compost are given names to encourage unsuspecting buyers to purchase them. They may have words like *black* or *humus* in them. Currently there are no standards or restrictions on what types of materials may be called compost. Example: Spent mushroom substrate (SMS) is called mushroom compost for marketing purposes. Many of the starting materials are the same as those used in composting. Often, enough salt (NaCl) is added so nothing can live in the substrate except the species of desired mushrooms. Once the mushrooms have removed the nutrients and yields are declining, the substrate is replaced with fresh material. The old material must be disposed of, so it is given the name mushroom compost to encourage people to purchase it. (See http://www.natureswayresources.com/resource/infosheets/mushroomcompost.html for more information.)

Screening: Screening is an extra step some producers use to ensure more uniform and higher-quality compost. Material larger than 1 inch in diameter is generally not fully composted and not ready to use, so screening will remove this material. If a person wants to apply compost to a lawn or other turf area, it needs to be screened small enough that the compost will filter down between the blades of grass. Fine-screened compost is also required for the machinery that fills potting flats in a nursery. For both of these usages, ⅜-inch screened compost is typical. *Note:* It costs a lot more to screen material to smaller sizes.

Quality Testing: Ask for the Solvita Compost Maturity Test and for test reports. Producers who are concerned with value and quality at least run chemical analyses of their products. Vendors who value quality and service will run biological test analyses in addition to the chemical. The biological reports are far more valuable in determining compost usage and quality. A mature compost will have a C:N ratio of 15–25:1. Buy from a reputable dealer and know what you are getting.

Regional Issues: The landfill dump rates are very high in some states, such as Minnesota, where they average $58 per ton, and may be much higher in some areas. An operator can therefore sell compost very inexpensively and still be profitable. Texas and much of the South has some of the lowest landfill dump rates in the country, so the cost of processing and any profit has to come from the sale of the product, not from waste disposal fees, hence compost prices are much higher.

Texture: Good compost is earthy smelling like the forest floor, loose and crumbly when you feel it, and slightly moist. If it is too dry, then many of the good microbes will have died or gone dormant.

Vermicomposting

Vermicomposting is the process by which earthworms are used to turn organic waste into very rich compost that is composed of earthworm castings (manure). Earthworms ingest the raw organic material along with soil and any microbes on it. The microbes living in the gut of the earthworm break down the material into a humus-rich substance that is called vermicompost. As the material is broken down, the nutrients contained in the organic matter are concentrated in the castings and in the microbe bodies living in the castings.

A variety of worms have been used in vermicomposting; the most common is *Eisenia foetida* (the tiger or brandling worm). Other suitable species include *Lumbricus rubellus* (the redworm), *Eudrilus eugeniae* (African night crawler), and *Perionyx excavatus* (Asian species). Each species has particular favorable characteristics, so it is important to choose the right one for the feedstocks being composted.

Vermicompost is most often produced by placing the worms and the feedstock into some form of bin or container. Open piles similar to windrows have been used but are not as common. The material must be kept moist and cool, as worms are very sensitive to environmental conditions. The processing of organic wastes by the earthworms is most rapid if environmental conditions have a temperature of 60°F–79°F (15°C–25°C) and a 70–90 percent moisture content. Depending on the species of worms used, they can daily process between 25 and 50 percent of their weight in feedstock material. Earthworms are very sensitive to salts and many chemicals, thus precomposting of wastes can disperse salts and ammonia, which are toxic to worms. A good example of the use of worms is for biosolids, which worms can turn into a more usable product with excellent pathogen reduction. (See Information Resources for many good books and papers on earthworms and vermicomposting. A new book, *Vermiculture Technology*, is an excellent source of information.)

Benefits of Earthworms: Earthworms improve the physical structure of the soil, which improves water infiltration and absorption rates, helping the soil drain better. Less runoff equals less watering and less erosion. The earthworms' tunneling activity improves soil aeration, porosity, and permeability. The increased moisture absorption results in more moisture being available to plants. The castings (the material that passes through earthworm bodies) absorb water faster than soil does and hold more water than equivalent amounts of soil. Castings also have the ability to absorb moisture from the air and hold it in a manner that plants can use. One study found that with good food sources and favorable conditions, a field might have over one hundred night crawlers per square yard. These deep-tunneling worms facilitate the movement of water into the water table, thus reducing runoff. Additionally, although the exact mechanism is unknown, studies at Ohio State University have shown that human pathogens (*E. coli*, *Salmonella*, cholera, etc.) that just come into contact with the skin of an earthworm are killed.

A healthy soil will have twenty-five earthworms per square foot of soil, which is equal to one million earthworms per acre. Studies in England have shown that in healthy soil, forty tons of castings are produced per acre. One study done in the United States indicates that 12 million worms per acre will move 20 tons of earth each year.

The tunneling activity of worms helps break up hardpan and compacted soils. Studies have shown that 30 percent of a field's respiration during cold, wet winter–spring months is due to earthworms. A European study found that in orchards, earthworms could increase the pore space in soil by 75–100 percent and that earthworm burrows accounted for two-thirds of a soil's air-filled pore space (see *Earthworm Ecology and Biogeography in North America*, 1995, in Information Resources).

Earthworms improve soil fertility by bringing up minerals from deep in the subsurface that are often in short supply in surface layers. The earthworm activity counteracts leaching by bringing up nutrients from deep in the soil and depositing them on the soil's surface as castings. The earthworm tunnels allow oxygen in the air to move deeper into the soil, which allows roots to go deeper and reach additional nutrients and moisture. As earthworms remove litter from the soil surface by ingesting it, they leave the nutrients and microbes in their castings for plants to use as a natural nonpolluting fertilizer. The microbes in earthworms' guts help destroy harmful chemicals and break down other organic wastes.

Earthworms create fertile root channels, as the mucus lining of abandoned burrows is an excellent source of nutrients and plant growth hormones. Plant growth stimulants such as auxins are produced in the worm castings; these hormones stimulate roots to grow faster and deeper. Additionally, earthworms chelate nutrients, making minerals available to plants that would otherwise be in a form that would make them chemically unavailable. Earthworms stimulate beneficial microbial populations. Nitrogen-fixing bacteria are more numerous near earthworm burrows and in their castings. One study on bacteria and actinomycetes found densities from ten to one thousand times greater (see *Earthworm Ecology and Biogeography in North America*, 1995).

Worms neutralize soil pH, for chemical analysis shows that the product coming out of the back end of a worm is closer to neutral than what goes in the front end. Analysis of earthworm castings reveals that they are richer in nutrients than surrounding soil, often have three times more calcium and several times more nitrogen, phosphorus, and potassium (see "The Abundance of Earthworms in Agricultural Land and Their Possible Significance in Agriculture"). Nitrogen-fixing bacteria that live in the guts of earthworms and in earthworm casts have higher nitrogenase activity, meaning greater rates of nitrogen fixation, which in turn is found in casts as compared to surrounding soil. One study found that earthworms are responsible for passing nitrogen to the soil at a rate of 100 kg N per hectare per year (see *Earthworm Ecology and Biogeography in North America*).

Earthworms improve plant growth and health. Tests have shown that crops grown in earthworm-inhabited soil increased yields from 25 percent to over 300 percent more than in earthworm-free soil ("Abundance of Earthworms"). Earthworms help eliminate thatch in lawns and grassy areas by eating and digesting the plant debris. Studies have shown that soils rich in earthworms have fewer harmful nematodes. Earthworms create soil conditions that discourage populations of soil organisms such as detrimental insects, parasitic nematodes, and others that are harmful to plants ("Vermicomposts Suppress Plant Pests and Disease Attacks"). One study found that earthworm-produced compost (vermicompost) dramatically increases germination and growth in many plants. Adding only 5 percent vermicompost to commercial growing media (95%) significantly increased plant growth (Clive A. Edwards, Ohio State University, *Nursery Management & Production*, January 1995). Research has shown that twice as many roots grew in pure worm castings than in sphagnum peat moss (Clive Edwards, Ohio State University). Experiments at Tennessee Technological University found that 10 percent vermicompost in a potting mix improved the germination of seeds of low viability (*Echinacea purpurea*) by 43 percent.

Studies have also found that many species of earthworms actually eat plant pathogens and in the process they also increase beneficial microbes. For example, a study found that in feeding, earthworms consume spores of mycorrhizae, beneficial fungi that help roots take up nutrients and water. These spores are deposited in the worm castings, deep in their burrows, where roots find them as they grow.

Other studies have shown that earthworms can increase yields of barley by 78–96 percent, spring wheat and grass by 400 percent, clover by 1,000 percent, and peas and oats by 70 percent. Research has also shown that yields were increased for millet, soybeans, lima beans, and hay. Studies in New Zealand found that earthworms at least doubled yields in all cases, and adding worms to crops has become standard agricultural practice there (*Earthworm Ecology and Biogeography in North America*, 100). Researchers at Oregon State University have found that a tea made from the worm castings speeds up the sprouting of hard-to-germinate seeds following a one-hour soaking.

A large earthworm population suppresses weed growth. The tunneling activity of earthworms prevents many of the conditions that weed seeds need to germinate. Earthworms often eat weed seeds and either destroy them or reduce their ability to germinate. Earthworms stimulate the growth of microorganisms in the soil, and some weed seeds are destroyed by these microorganisms. Some microorganisms (bacteria and fungi whose growth is stimulated by worms) live in a symbiotic relationship with plant roots and help plants grow better, thus shading out weeds and outcompeting them for water and nutrients.

Worms often help clean up dangerous chemicals in the environment. Researchers have found that bacteria living in the guts of worms break down (detoxify) many hazardous chemicals, such as hexachlorocyclohexane

(HCH). Microbes living in worms have the ability to break down complex organic molecules like cellulose and lignin.

Earthworms improve water absorption and prevent erosion. The castings increase the water stability of the soil, because earthworm castings can take a direct hit by a raindrop and maintain their shape, which reduces erosion and runoff and helps the soil absorb water. A research study conducted in Minnesota showed that earthworms added to cornfields increased water absorption rates thirty-five times over control fields without the earthworms within a six-week period. In soil in a field with one hundred night crawlers per square yard, 2 inches of water (a very heavy rainfall) could be absorbed by the soil in twelve minutes. The same soil without earthworms took over twelve hours to absorb that much water (USDA National Soil Tilth Lab). Another study found that if the top 3 feet of soil contained 25 percent macropores (earthworm burrows), then that soil should be able to absorb at least a 9-inch rainfall without runoff. Note that tilling the soil not only kills beneficial fungi but reduces the number of earthworms, which affects runoff. One study showed that on a sloping field with no-till practices, there were 155 earthworm holes per square yard and an average runoff of 0.08 inches per year. This compares to a tilled field with 6 holes per square yard and 4.9 inches of runoff per year (the average rainfall for this area is 39.4 inches). Scientists from the Agricultural Research Service found that grass- and leaf-mulched plots had twice as many earthworms as those mulched with cornstalks. Water penetrated the earthworm-filled soil up to four times faster.

Some scientists now believe that earthworms have the potential to eliminate soil erosion! This could save society billions of dollars in erosion control, reduce pollution from dangerous synthetic chemicals, and improve the environment.

Research presented at the ISEE 5 (International Symposium on Earthworm Ecology at Ohio State University) points to earthworms being an important biomedical resource. It has been found that ingredients from earthworms have anticancer properties. The bodies of earthworms are extremely nutrient rich in minerals, amino acids, proteins, and vitamins. When earthworms die, these nutrients are released into the soil. This is also why they are a good animal feed.

To attract and promote earthworms, the soil needs to be mulched with organic mulches that help stabilize soil temperature and moisture. The mulch provides food and shelter for earthworms. Native mulch and compost are the most valuable for promoting earthworms. The increased particle surface area of the small particle sizes in native mulches that have been ground also allows for the greater microbial activity that is preferred by worms. Rough (unfinished) compost is one of the best worm-food mulches there is. Studies have found that most organic fertilizers tend to have a positive effect on earthworms and increase population densities (*Earthworm Ecology and Biogeography in North America*). Most artificial synthetic fertilizers reduce earthworm populations.

Earthworm Damage by Synthetic Chemicals

Agricultural chemicals such as salt-based artificial fertilizers (i.e., 13–13–13), pesticides, etc., can kill earthworms. Even if a few pesticides such as DDT do not kill earthworms, birds are killed when they eat the worms (*Pesticide Reviews* 57, 1975). Earthworms and other beneficial organisms are destroyed by synthetic chemical fertilizers and fungicides, pesticides, etc. (*Reviews of Environmental Contamination and Toxicology,* 1992). In the absence of earthworms, the soil becomes lifeless, sterile, and nutrient deficient.

Types of Earthworms

Over three thousand worm species have been identified. Experts disagree as to what distinguishes one type of worm from another and if one species is a true earthworm or not. All soil worms are beneficial, and most references lump all soil worms into the category of "earthworms." There are two basic types of worms: those that feed on the surface and those that feed in the subsurface. The surface feeders eat plant residue, are generally large worms, and live in vertical burrows often over 6 feet deep. Subsurface feeders are smaller than surface feeders like night crawlers but outnumber them nine to one. They eat their way through the subsurface, loosening, aerating, and improving soil structure in the process.

When worms are separated into "worms" and "earthworms," the following distinctions apply. Redworms, often called manure worms, brandling worms, or red wigglers, are reddish brown in color and live in the soil in the surface layer of decaying vegetation (litter). They feed on this layer, multiplying rapidly in numbers, expand into poorer surrounding soil, and die, thereby distributing the nutrients contained in the excess wastes over a larger area. They are often used in small-scale worm bins. *Eisenia foetida* and *Lumbricus rubellus* (which tends to be more soil dwelling if large amounts of organic material are in the soil) are examples of redworm species. Earthworms, often called soil-processing worms, are burrowers and soil processors that eat dead organics and rock particles, grinding and excreting them as a finely ground mix that serves as food for bacteria. They tend to survive in harsh conditions better than redworms. They do not assimilate the organics for themselves to the same extent as redworms, hence they do not multiple as quickly as redworms, whose assimilation rates are much higher. The redworms' higher rate of assimilation means that the nutrients they consume go into building their own biomass, whereas earthworms pass on these nutrients in a soluble form in their castings. *Pheretima elongata* is a deep-burrowing earthworm used in Mumbai, India, to convert garbage into vermicompost that is recommended by the Bhawalker Earthworm Research Institute as the most efficient organic waste converter. Waste conversion occurs at the soil surface, not in a bin, thus less material handling is required. *Lumbricus terrestris*, called night crawlers, dew worms, rain worms, or orchard worms, like soil temperatures less than 50°F. They also dig burrows and do not like to have their burrows disturbed. They come to the surface to feed on dead grass, leaves, etc., drawing the organic matter deep into the soil layer. Other garden worms are *Allolobophora caliginosa*, *A. chloritica*, *Aporrectodea turgida*, and *A. tuberculata*—all often found in pastures. Most worms found in U.S. soils are not native.

Using Compost

Flower Bed Preparation: A layer of compost 3–4 inches thick should be tilled or turned thoroughly into the top 4–8 inches of soil for most plants. Organic fertilizers and other amendments can be added at this time. Shallow-rooted plants like bedding plants for annual color require less, about 2–3 inches of compost turned into the top 4 inches of soil. For best results, top-dress the beds with 1–3 inches of compost or 1–3 inches of composted native mulch every year.

Lawn Preparation: For new lawns, apply a minimum of 1 inch of compost (2–3 inches will help grass become established quicker and require less water) tilled into the top 4 inches of soil before sod or seeds are planted. Some will argue that applying a thin layer (½ inch) of compost on top of the sod or establishing seedlings is a better approach. Organic fertilizers, trace minerals, and other amendments can also be added at this time. Each year after the turfgrass is established, top-dress the lawns with ½ inch of finely screened compost. Many turf managers like to use ¼ inch of compost applied twice a year.

Orchard and Fruit Trees: Apply coarse compost 2–3 inches thick under each tree, starting about 1 to 2 feet away from the trunk and extending well beyond the drip line. An annual compost application should be between ½ and 1 inch thick.

Herb Gardens: Mix a 3-inch layer of compost into the soil before planting or mulch existing plants with 2 inches anytime.

Grapes and Berries: Grapes and berries have shallow root systems, so cultivating around them can be troublesome. A good application of coarse compost 3 inches deep in early spring will help suppress weeds, reduce the need to cultivate, and feed the plants.

Flowers: Compost may be safely applied to all flowers. Work it into the soil or use it to mulch the surface. When used on the surface, some people like to cover the compost with a shredded coarse-textured mulch to keep it from washing away. Note, however, that once the compost has been watered in and in place for a while, it is very resistant to washing out.

Vegetable Gardens: For most vegetables, a mix of two parts soil to one part compost works well. After planting transplants or after seed germination, a good mulch layer 2–3 inches thick should be added. Many people use compost for the mulch layer.

Compost Tea

A good compost tea has been shown to help prevent or control a wide range of foliar diseases in greenhouses, field crops, fruit trees, ornamental crops, and turfgrasses. Compost tea is now being used for soil disease control by injecting it into central pivot and other irrigation systems. It often works better, is far cheaper than toxic fungicides, and does not have the negative environmental impacts.

Compost tea is now being used by many professionals: nursery owners, turf managers, golf course superintendents, landscapers, ranchers, farmers, and recreation departments, as well as in orchards and municipal parks. Anyone involved with the growing or maintenance of plants can make use of its benefits.

History

There are many types of products called compost tea, but it can be broken into two distinct types, aerobic and anaerobic. The study of compost tea started with farmers and gardeners placing manure into a bucket of water and letting it sit for 7–10 days and then watering plants with it (manure tea). Sometimes they observed increased growth and disease reduction, and other times it did not work and pathogens were present. Researchers at universities in Israel and later in Germany started developing methods to increase the consistency in both composting time and efficacy. These teas were really extracts, since they were made by adding finished, properly prepared compost to water and stirring the mix for 1–14 days at room temperature.

Researchers started experimenting with ways to increase the microbial density and diversity of the extract by adding food for the microbes, such as molasses, sugars, fish emulsions, etc. The microbe populations grew so high and so fast that the solution became anaerobic and started fermenting, which allowed pathogens to grow and created odors. This research led to aerating the tea, which produced more consistent results. Over time, researchers found that actively aerating the tea and adding microbe foods gave even better results. During the same time span, microbiologists were developing better tools to study microbes in soils and compost and how they relate to plant health and growth. The model of soil fertility and management called the Soil Food Web that resulted is covered in Chapter 2. Our knowledge of how to produce and use compost tea is continually growing.

The easiest and most consistent methods became known as Aerated Compost Tea (ACT) and more recently Actively Aerated Compost Tea (AACT), which implies a tea made with a focus on stirring and aeration. The basic idea is simple: Compost is placed in water, and air (oxygen) is added to the liquid mixture, usually by bubbling air into the system, which also serves to stir and mix the tea. Microbes are washed out of the compost into the water and allowed to multiply to very high levels. If the tea is properly made, it will have a mix of facultative and aerobic microbes at very high densities. It is very important to note that the tea will only be as good as the starting materials (i.e., the compost used), so always use the best compost you can find. A producer of quality compost will have microbial test reports available indicating the amount of bacteria, fungi, protozoa, and nematodes in the compost.

Some researchers have taken animal manure (cow patties) and placed it in a bucket of water for a few days, allegedly applied it, analyzed the results, and concluded that compost tea does not work. This is not compost tea but manure tea. Similarly, other researchers have taken items like alfalfa pellets and soaked them in water for a few days and concluded that compost tea does not work. This was alfalfa tea or extract, not compost tea. Another product sometimes erroneously called compost tea is "compost leachate," which is the liquid that drains out of compost when it is overwatered and flows out of the bottom of the piles. Compost leachate is not compost tea but still has beneficial properties. Compost "tea" can also be made by extraction, whereby the beneficial properties are pulled from the compost and sprayed immediately without any brewing at all.

Note: The U.S. National Organic Program (NOP) has strict guidelines on the use of compost teas in USDA-certified organic farming.

Mode of Action

Compost tea is sprayed onto the surface of leaves and stems of plants. Bacteria such as *Bacillus* and *Serratia* and fungi like *Penicillum* and *Trichoderma*, along with many others (actinomycetes, yeasts, protozoa, nematodes, etc.), are the main agents. The microbes act in several main ways: by inhibition of pathogen spore germination, through antagonism by competition with pathogens, and through induced host plant resistance against pathogens.

The best teas are brewed for twelve to twenty-four hours and used immediately or at least within twenty-four hours if the tea is kept aerated and not allowed to go anaerobic. The beneficial microbes in the tea are living creatures that require air (oxygen) to live. Without an air source (constant aeration), they will die from asphyxiation, so beware of products called compost tea sold in closed bottles.

Compost Tea Brewers

Brewers for compost tea can be as simple as a 5-gallon bucket or as complex as a very expensive machine costing thousands of dollars. It all depends on the quality and quantity of tea required.

One type of machine to produce a form of compost tea is called an extractor. The tea produced is called compost extract or liquid compost. Most brewers only remove a small portion of the microbes from the compost and then add food and air over time to allow the microbes to reproduce to the high densities required for good results. An extractor removes many times more microbes from the compost to get the high densities and can produce large volumes of tea very quickly, from a few hundred gallons an hour to over two thousand gallons per hour for large extractors. The extraction process tends to disturb the microbes and put them into their spore or dormant form. Hence, a compost extract may have a shelf life of four to ten days instead of twenty-four hours, which is an advantage in some commercial operations. Microbial foods are added shortly before application to wake up the microbes and get them active and growing.

Compost Tea Types

From the soil food web fertility model, it is known that some plants like fungi-dominated soils while others prefer bacteria-dominated soil. By adding the correct foods to the brewing cycle, one can steer the tea to be dominated by bacteria or fungi. An example usage of this concept is for weed control in pastures. Most of the perennial grasses that offer the most protein and benefit for livestock prefer balanced to fungi-dominated soils. Most of the common weeds that ruin pastures prefer bacteria. Artificial fertilizers, tillage, compaction, and overgrazing make the soil bacteria-dominated, which favors weed growth. By adding fungal foods (fish emulsion, humate, etc.) to our compost tea, we can make it very fungal. Over time as this tea is applied to pastures, the weeds die off and the amount and quality of the desired grasses increase as the natural balance is restored.

Getting Started

A simple brewer can be made from a 5-gallon bucket or purchased from many sources. A few years ago, Elaine Ingham of Soil Foodweb, Inc., tested many brewers using the same compost as a starting point. The second-highest-quality tea was produced by a homemade design. Follow the steps in Appendix 1 to make compost tea, whether in a small bucket, a large commercial machine, or a homemade brewer of any size. The basic idea can be scaled up to a 55-gallon drum, 250-gallon tote tank, or even larger if greater quantities of tea are required.

Diseases Controlled with Compost Tea (a sample list)

Alternaria solani: early blight on tomatoes
Botrytis cinerea: gray mold on strawberries, geraniums, beans, tomatoes, peppers
Diplocarpon rosae: black spot on roses
Pseudomonas syringae pv. maculicola: leaf spot
Sphaerotheca pannosa var. *rosae*: powdery mildew on roses
Sphaerotheca fuliginea: powdery mildew on cucurbits
Uncinula necator: powdery mildew on grapes
Venturia inaequalis: apple scab
Venturia conidia: apple scab
Xanthomonas vesicatoria: leaf spot on tomatoes
Podosphaera pannosa var. *pannosa*: powdery mildew on roses
Pythium ultimum: damping-off in cucumbers
Fusarium spp.: root rot in cyclamens
Fusarium oxysporum: fusarium wilt
Monilinia laxa: blossom rot on cherries
Phytophthora spp.: root rot on avocados
Phytophthora infestans: leaf blight on tomatoes and potatoes

Vermicompost Tea

This tea is similar to regular compost tea except vermicompost (earthworm castings) is used instead of thermal compost. Since earthworms process the organic material into vermicompost (VC) at room temperature, we get a different mix of microbes that cannot live in the high temperatures in regular compost. Vermicompost teas (VCT) have been shown to increase germination rates of seeds, increase the growth rate of plants, and provide disease control.

The VCT contains nutrients, humic and fulmic acids, hormones, growth stimulators like auxins, plant growth regulators, and microbes. These are believed to be the major contributors to the measured benefits. Researchers have found that the humic acids from VC work better than standard commercial humic acids even though the mechanism is unknown at this time.

VCT has been found to prevent and treat *Pythium* on radishes, *Rhizoctonia* on cucumbers, verticillium wilt on strawberries, phomopsis and powdery mildew on grapes, bacterial rot on cucumbers, and other common pathogens. *Note:* If the VC is sterilized before application, then one does not get the benefits, clearly demonstrating that the effect is microbial in nature.

Most compost tea is now brewed using a mix of 80–90 percent thermal compost with 10–20 percent vermicompost, as this combination has been shown to give the best and most consistent results.

Mushroom Compost, or SMS

First, mushroom compost is not real compost. The real name of this product is spent mushroom substrate (SMS). It is called "mushroom compost" as a marketing ploy to help dispose of it. The quality and usage of SMS varies greatly around the world, from good and useful to very bad and toxic, based on regional customs and local regulations.

SMS is made from a combination of wheat straw, dried blood, gypsum, lime or crushed limestone, poultry litter, cow or horse manure and bedding, hay, corncobs, cottonseed hulls, cocoa bean hulls, clay, peat moss, cotton seed hulls, or other items, depending on what is available in a given area. The material is partially composted for a few weeks and then steam-pasteurized (sterilized), which kills off all of the beneficial microbes. Then the material is inoculated with the species of mushroom fungus that the grower wishes to produce. Additionally, the material is often loaded with table salt (sodium chloride, NaCl) to ensure that only the desired species of mushroom will live and grow in the substrate.

Several crops of mushrooms are raised and harvested on this substrate until yields start to decline as the original nutrients are used up. When the nutrient level drops too low to raise mushrooms, the substrate is replaced with fresh substrate. The used substrate must then be disposed of. Worldwide, this is over 4 million tons per year!

Common Problems of SMS

Studies have shown decreased plant growth and yield at levels as low as 5 percent SMS in a mix due to high soluble salts. Other reports have shown that it may have a high pH level, which is harmful to plants that like acidic conditions.

Even the low-quality bagged manure often sold in big box stores often has three times the amount of nutrients that SMS does.

Mushroom growers have major problems with fungus gnats, so they regularly spray with toxic chemicals such as methoprene, cyromazine, diflubenzuron, Dimilin, and diazinon. Other toxic chemicals occasionally used are benomyl, thiabendazole, and chlorothalonil. For this reason, SMS does not meet the standards for use in organic production.

SMS is often stored in large piles that become anaerobic. This allows pathogens to grow in the material. The putrefying organic matter creates organic acids that often have a strong odor. Common odors are vinegar, sour milk, vomit, rotting meat, and occasionally ammonia or rotten eggs. All these odors indicate the presence of alcohol, which is toxic to plant roots in concentrations as low as 1 ppm.

Benefits of SMS

SMS is organic matter and still contains some nutrients. Since most of our soils are very low in organic matter, it may provide some benefit. Best results occur in sandy soils in areas with lots of rainfall so the water can wash out and leach the salts. *Note:* In clay soils, the salts help glue the particles together, thus helping create hardpan.

Killer Composts

Most states do not have regulations governing the type or quality of products called compost. As a result, many so-called compost products range from quite poor to even very bad for plants. These "killer composts" are often caused by common herbicides.

Washington State University was the first to discover the link between persistent herbicides and plant kills caused by compost. Other kills have been confirmed in almost every state (Washington, California, Pennsylvania,

Texas, Maine, etc.) and in many other countries such as New Zealand, Great Britain, and many more.

Damage shows up as cupped leaves, distorted fernlike foliage, pale color, distorted growth with prominent veining, stunted growth that does not produce crops, and death of the plant. Most recently (2008), thousands of gardens in Great Britain have been destroyed by compost containing the herbicide aminopyralid.

Many types of herbicides are used to treat our lawns, pastures, or hay fields for weeds. The grass clippings and hay are often used in making compost. Grass, hay, and even grain are also used as animal feed, then the manure and bedding are used to make compost.

Many of these herbicides are so toxic and persistent in the environment that a horse or a cow can eat the treated hay or grain, it can then pass through the animal's digestive system and be excreted in the manure and urine, be absorbed in the bedding, be collected and then composted, and still kill plants!

The breakdown rate of an herbicide is often expressed as the half-life (the number of days it takes for 50 percent of the chemical to be degraded or broken down into harmless components). If an herbicide breaks down quickly, it must be applied more often, hence the chemical companies try to make the herbicide more persistent. If an herbicide is more persistent, then when it is used, applicators do not have to apply it as often, saving them money. This is a great marketing tool and justifies a higher price for the herbicide product (i.e., profit).

If an herbicide has a half-life of thirty days or less in the soil, it is considered nonpersistent, whereas herbicides that have a half-life of over one hundred days are considered persistent.

Persistent Herbicides

Aminopyralid: Released in 2005 for use on pastures to control perennial weeds; has been associated with the loss of thousands of home gardens in Great Britain in 2008 and more recently in Washington State. It is found in the products Milestone, CleanWave, and ForeFront. It has a reported half-life of 533 days in the soil and is now being used in the United States. Broad-leaved vegetables and plants are the most susceptible (e.g., roses, tomatoes, potatoes, lettuce, carrots, beans, and peas).

Clopyralid: Found in the products Tordon, Confront, Lontrel, Stinger, Transline, TruPower, Millennium, and Millennium Ultra. It can cause severe eye damage, including permanent loss of vision, in humans. The active ingredient in Confront has been found in compost that killed home and nursery plants in Washington, Pennsylvania, Texas, New Zealand, and many more places. It is extremely toxic to composites (sunflowers, marigolds, lettuce), legume crops (beans and peas), and nightshades (eggplant, peppers, tomatoes, and potatoes). It will cause damage at levels fifty thousand times lower than that allowed on grass. Even after one year of composting, levels have been measured that are ten to one hundred times greater than is required to harm sensitive plants.

Triclopyr: Found in the products Garlon, Turflon, Access, Redeem, Release, Weed-B-Gon, Crossbow, Grazon, ET, and many others. It is resistant to decomposition, with a half-life of 30–90 days in soil. However, the half-life of the breakdown products ranged from 8 to 279 days. It inhibits the growth of many beneficial fungi such as the mycorrhizal and prevents nitrogen fixation by microbes.

Picloram: Used to kill weeds on pasture and rangeland, railroad tracts, easements along power lines; is water soluble and very mobile. One study found that it may take several years to degrade in the soil. It has been found to be the active ingredient in many toxic composts from around the country. Found in many herbicide products.

Simazine: Found in the products Simazina, Atanor, Gesatop, Princep, Caliber 90, Simazine, and many others. Extremely resistant to decomposition, and 50 percent has been found in soil up to two years after application, depending on soil type. It has been banned in Europe.

2,4–D: Used in over fifteen hundred products including Agent Orange. It does break down after thorough composting, but over one year is required.

Herbicide Breakdown

The breakdown rate of herbicides in compost is influenced by many factors. These include the starting materials (feedstocks), the composting techniques or process used, time factor (length of composting), moisture content, and so on.

For herbicides to break down in a compost pile, several things are required:

1. Microbes found in anaerobic conditions (without oxygen) as well as microbes found in aerobic conditions (with oxygen)
2. Various fungus species
3. Long periods of time (enough time for multiple half-lives to occur).

Composters that use the static pile method will compost the material from six months to over three years! This allows for many half-life cycles to occur, breaking the herbicides down into harmless components.

These persistent toxic herbicides are often used on pastureland, hay fields, grain crops, and grasses. They most commonly occur in compost when cattle are fed hay or silage (grain) from fields that were sprayed with herbicides for weed control. The animal bedding and cow manure become saturated with these herbicides, the manure and bedding are then composted, and we now have killer composts. The highest herbicide levels are found in compost made from hay, grass clippings, cow manure and bedding, and poultry manure and bedding.

Gardeners should be wary of buying the low-cost composts that contain cow or poultry manure, as the chance of encountering herbicide residues and the resulting problems greatly increases. Ultimately, we are responsible for what we put on our yards and gardens . . . and when one buys something cheap, they get what they are paying for.

CHAPTER 7

Mulch

Walk into the woods or onto the prairies and look around; you will be in the presence of much life—plant and animal, large and small. Then look down; you will see an equal amount of death, many expired life-forms covering the soil. You will find a mulch of dead things—twigs, leaves, grass, insects, manure, and even dead animals. Dig into the mulch, and you will find it beginning to decay—compost. The deeper you dig, the more advanced the decay, until individual pieces fade into rich, moist topsoil. Topsoil is the digestive system of the earth. It keeps the water and air clean and furnishes the food for all life. The quality of all life on earth—including human life—depends on the quality of the topsoil. That crucial thin layer of soil must be protected, maintained, built, and nourished. The mulch cover of organic materials performs this service and much more.

Mulch Types

To be a successful gardener, grower, or horticulturist, one must mulch. Mulch comes from the German *molsch*, meaning "soft," and refers to any loose, generally soft material that is laid down on top of the soil to protect a plant's roots or spread lightly over the plant itself.

Mulch is not a soil amendment; it is a covering or surface layer used to protect the topsoil. Nature does not allow bare ground, so neither should we. *Mulching is considered to be the most important step in any gardening program.* All natural or organic mulches will improve the soil, but like all things, they vary in quality and effectiveness. Good mulch lets air (oxygen) and water enter the soil and allows carbon dioxide to escape. Good mulch will readily decompose, releasing the stored nutrients, and will provide microorganisms and earthworms a good home and food source.

Mulches add to the aesthetic value of a garden while protecting the bases of shrubs and trees from injury by mechanical equipment. Mulches reduce the energy reflected onto the walls of your house, lowering the amount of energy required to cool it (less energy used means less pollution is generated in its production). Mulches prevent soil compaction and provide a home for beneficial insects and animals. Mulches are so important, in fact, that in 1989 the state of California passed a water conservation bill that, among other things, required the use of mulches!

Shredded native tree trimmings mulch—the best mulch in many ways. Photograph by Howard Garrett.

For years, mulches were divided into two basic types, organic and inorganic, of which organic mulches are the most valuable. Mulches, like many other products, range from excellent to very poor except in one important aspect, the highest-quality and most beneficial mulches often cost less. Recently scientists have begun to recognize a third category of mulches called living mulches, and studies of their usage and benefits are now being published. Today's gardener has access to more types of mulches than ever before. Now the questions are which mulch to use and when, where, and how to use it.

Organic Mulches are composed of material that was once alive but is now dead and that will decompose or rot over time. Organic mulches may be totally natural (unprocessed), like leaves or pine needles, or they may be processed, like newspaper.

Inorganic Mulches are made of materials that were never alive and are generally inert. Materials like stone, crushed rock, plastic sheeting, glass, and so on are inorganic mulches. Except for specific cases and locations, inorganic mulches do not allow for natural systems and cycles to work, hence gardeners who use them will tend to have more disease and pest problems.

Living Mulches are a new class of mulches that are composed of living plants that are planted for a specific purpose in covering the soil. This term is also used for mulches that are made up of compost and shredded tree trimmings. Living mulches are an extension of an agricultural practice called cover-cropping.

Organic Mulches

Organic mulches offer the most benefits, often at lower cost, and improve the fertility and health of the soil. A 3-inch layer of organic mulch can lower soil temperature by about 25–30°F, which reduces plant stress and water requirements. Bare soil can easily reach 100°F–135°F, which speeds evaporation and dries out the soil, which stresses the plant, resulting in wilting, more insect and disease problems, and in most plants, eventual death. The higher soil temperature reduces plants' roots ability to absorb moisture (even if it is there) and kills beneficial microbes that help feed, water, and protect plants' roots. In addition, soil nitrogen (N) decreases as soil temperature increases. For every 10°C increase in soil temperature, soil nitrogen will decrease two to three times.

Studies in Austin and San Antonio, Texas, during the 1990s found that homeowners and commercial property owners who mulched their lawns with ½ inch of compost each year saved $50–$200 per month on their water bill. Two studies from Ohio State University have confirmed that plants grown in organically enriched soil suffer far fewer disease and insect problems than those grown with synthetic chemicals. Hence good organic mulch helps build up the soil, naturally increasing a plant's pest and disease resistance. USDA studies on several species have found that mulched plants were often three times as large and produced three times the yield of unmulched plants after several years.

Bark Mulches

The most common type of organic mulch is bark mulch. Bark mulches are made from the protective outer layer of trees and are produced as a by-product of the lumber and pulp industries. Since outer bark is designed as a protective layer for the tree, it tends to be low in nutrients. Tree barks frequently contain the chemical suberin, a naturally occurring substance that waterproofs (helps bark shed water) and prevents the bark from being broken down by soil microorganisms. In addition to suberin, barks contain waxes that also help waterproof the tree. Hence, the suberin in the bark can slow or retard the growth of some plant species. Additionally, barks contain very few energy-releasing compounds used by the soil microorganisms that are extremely important to soil and plant health.

Barks can be broken into two basic types, hardwoods and softwoods (conifers). In much of the country, hardwood bark is mostly from oak trees, and softwood bark is from pine trees or other conifers. They are both a by-product of the lumber and paper industry. Since conifers tend to be a pioneer species (meaning they grow on poor, nutrient-deficient soils), they contain very few nutrients (less than hardwood bark). Barks have a very high C:N ratio that averages 450:1, so they require a lot of nitrogen to break down, often starving nearby plants in the process.

Large pieces of bark are slower to break down and less likely to blow and wash away than finely ground pieces, but they are considered more difficult to work around. Barks and uncomposted sawdust from redwood, cedar, Douglas fir, larch, eucalyptus, and spruce trees are considered toxic to many plants. Any bark that is high in tannic acids and phenols is potentially harmful unless thoroughly composted and leached.

Fine-ground pine bark packs down and prevents oxygen from reaching the soil. It is difficult to wet, sheds rain after it dries out, and prevents moisture from reaching the plants' roots. Often, when trees are dying from disease, they are cut for pulp or lumber and the diseased bark ends up being sold to consumers.

Research at Cornell University has shown that conifer barks release toxic volatile compounds that are harmful to plants like tomatoes. Research at the University of Arkansas has found that marigold growth was significantly reduced in beds mulched with pine bark. Many tree biologists, plant anatomists, arborists, soil ecologists, and other experts now recommend that bark-based mulches be avoided. Alex Shigo, a leading tree expert at the University of Georgia, has posted several papers on the Internet about this subject. Research at the Ohio Agricultural Research and Development Center has also found that pine bark does not support many of the beneficial microorganisms that prevent disease.

The natural chemicals in pine bark tend to kill off many species of beneficial microbes that naturally attack and prey on fire ants and termites; hence problems with these insects are reported to be more common when pine bark is used.

A report presented at a Texas Apple Growers Association convention found that bark mulches actually steal nutrients away from plants as compared to mulches made from recycled tree trimmings and brush (i.e., native mulches). As fuel prices have increased, more and more bark is being burned for fuel, thus reducing its availability for the nursery and landscaping industries.

Cedar Mulches

Many users report that they have had good results with cedar mulches, while others report extreme failures and problems. This issue needs further research to explain the difference in results. Some patterns have been observed regarding mulches made from species with the common name of junipers (*Juniperus ashei*, *Juniperus deppeana*, etc.) versus those from species commonly called cedars, such as the eastern red cedar (*Juniperus virginiana*), southern red cedar (*Juniperus silicicola*), and western red cedar (*Thuja plicata*). The first group tends to grow on alkaline calcareous soils, while the second group is often found growing on more acidic soils. John has seen some negative results with cedar mulch in the Houston area. Howard has seen nothing but positive results in the Dallas–Fort Worth area.

A second issue with using cedar mulches is the type of processing they have had. This could range from fresh ground cedar from land-clearing operations to de-oiled cedar flakes from industry. When de-oiled cedar flakes from industry are used, they can weigh as little as 190 pounds/cubic yard, since they have been cooked at temperatures of 225°F or more to remove the oils and have almost zero moisture when they leave the mill. The low weight is good for bagging, handling, and shipping purposes. The cedar flakes can absorb water after application and weigh over 1,600 pounds/cubic yard wet. De-oiled flakes are very low in nutrients and may cause nitrogen deficiencies in the soil and other mineral tie-up problems (i.e., very high carbon:nitrogen ratio). If the cedar flakes are soaked with an organic fertilizer such as seaweed, fish emulsion, or a good compost tea, they can make a better mulch.

Fresh ground cedar may be used directly but is almost always more effective if composted for a while before using. Western red cedar (*Thuja plicata*) is very rot resistant. While this characteristic makes it good for lumber, the chemicals such as resins and oils that act as preservatives are not good for plants. These chemicals resist attacks by insects and microorganisms (e.g., fungi), hence also render the wood toxic for soil or composting organisms. Any mulch from this tree should be composted for a long time first to help break down these chemicals.

Note: Cedar from younger trees has less of these chemicals than older trees, which contain more heartwood and have accumulated more of these compounds.

Native Mulches

For years, gardening experts have claimed that this was the best mulch of all, and now scientific research is backing them up. "Native" mulch is made from recycled fresh green tree and brush material that was recently alive and comes directly from a grinding operation. The Texas Association of Nurserymen (TAN) recognized native mulches as a class of mulches separate from barks and other materials in their 1997 product directory. Native mulches are available as fresh ground or aged (composted) and come in many variations.

Native mulches started becoming available in recent years as society became aware of the importance of recycling brush and tree trimmings instead of burning them or placing them in a landfill. Recent research has found that mulches made from recycled native trees are the highest quality available. They are also among the lowest in cost, since they are made locally and do not have high transportation costs associated with them.

Local native mulch is produced from a mixture of native trees (primarily hardwoods), conifers, brush, and any other species growing in a given area with bark, wood, and sometimes leaves included.

Native mulches have a high percentage of buds, shoots, leaves, and cambium bark layers in them. These materials are rich in protein and other nutrients, which is why deer and other animals eat them as a food source. These native mulches are many times higher in nutrients than barks. Native mulches encourage biodiversity of beneficial microbes and earthworms in the soil and feed plants as they decompose.

Years ago, consumer awareness was the only negative working against native mulches, since their appearance is different from pine bark or shredded hardwood bark. However, as these mulches have become more available, this perception has changed. Studies and market acceptance have shown that most people prefer the native mulch, since it actually looks more natural than barks or other alternatives.

Composted or Aged Native Mulch

Native mulch that is aged or composted first before application is of the highest quality. The heat generated during the composting process kills any pathogens and weed seeds that might have been present. The composting process also concentrates the nutrients contained in the raw material and stabilizes nitrogen. Additionally, the composting process breaks down the lignin and cellulose contained in the raw material, rendering a less attractive home for termites and many pathogens after it is applied. The composting process allows very high levels of beneficial microbes to develop and grow in the mulch, increasing its value.

Screened, composted native mulch is also an excellent amendment to use in soil mixes, as it supplies energy to the soil (stored in its chemical makeup) in the right form for beneficial soil organisms to use. Grinding and screening (particle size) will determine the appropriate usage. A two-year study from Texas A&M University (TAMU) has found that native mulch and compost outperformed all other erosion control methods. It was also the lowest costing mulch! Research in Florida has confirmed TAMU's work. Research at the Ohio Agricultural Research and Development Center has found that plants grown in substrates rich in biodegradable organic matter (such as found in native mulches) support microorganisms that induce systemic resistance to disease (*American Nurseryman* 186, no. 7 [October 1, 1997]).

As a bonus for those in the South dealing with imported fire ants, using high-quality composted native mulch may reduce mound density. Many landscapers, gardeners, and others have observed and reported a reduction (not elimination) in the number and size of fire ant mounds in areas where composted native mulches were used. It is believed that the native mulches increase the density of organisms that attack and prey on the fire ants, reducing their numbers. This has not been confirmed by rigorous research and would be an interesting area of study for our universities.

In general, a 3–4-inch thick layer of mulch should be used on ornamental beds and a 4–6-inch layer around trees and shrubs. It comes in many formulations and sizes. It is sometimes blended with shredded hardwood bark to obtain a familiar appearance (though this lowers its quality) or with compost to increase the quality. It has been used as a potting media in container-grown plants or to root cuttings and often works better than bark for many species of plants.

The general benefits of native mulch are these: It's economical; the composting process concentrates nutrients and stabilizes nitrogen, and the heat kills weed seeds and pathogens; it improves plant and soil health; it sets up quickly, it's reported to prevent many plant diseases; it encourages microbial biodiversity in the soil; it's reported to increase tree and plant growth rates. Additionally, it is subject to lower freight costs and less transportation that causes pollution, hence is much more environmentally sensible. The use of native mulch also saves valuable landfill space and avoids air pollution from burning, since it is made from recycled materials.

Using native mulch also helps reduce greenhouse gases. When organic materials are placed in a landfill, they undergo anaerobic decomposition, producing methane, which contributes twenty-three times more to global climate change than carbon dioxide. Also, since native mulches are made from recycled materials, they qualify for points in the Sustainable Sites Initiative (future Leadership in Energy and Environmental Design [LEED] landscaping guidelines) ratings.

Note: Coarse-ground and unscreened composted (aged) native mulch works best from a physical, chemical, and biological perspective. However, a screened version is more cosmetically appealing and works better as a soil amendment. It is sometimes available in a double-ground form that looks similar to some shredded barks.

Freshly Ground Native Mulch

This mulch comes directly out of the grinder. It has not undergone any processing or screening. It tends to be inexpensive and useful in special applications. As for all native mulches, it is a mix of whatever species came into the mulch/composting recycling facility.

If one is not in a hurry, this mulch is one of the best ways to naturally break up heavy clay soils and suppress weeds. To suppress weeds, it is often applied 4–6 inches thick and sometimes rolled or watered down to help it stay in place better. The mulch smothers the existing plants, essentially killing them. This mulch becomes very active biologically, and the microbes will use the nitrogen stored in the dead and dying weeds to help break down the mulch. The microbes will also break apart clay particles, creating a looser soil (see Chapter 2).

Freshly ground native mulch is also used for temporary road beds, erosion control, soil improvement, garden paths, land reclamation, filtration of storm water runoff, and any other application where large volumes are required and cost is an issue.

Wood Chip Mulches (single species)

Sometimes recyclers will grind up one species of tree into chips that can be used as mulch. Research has shown that these types of mulches may retard the growth of some species of plants. One study by Bartlett Tree Laboratory, UK, compared six species of pure wood chip mulch of beech (*Fagus sylvatica*), hawthorn (*Crataegus monogyna*), silver birch (*Betula pendula*), cherry (*Prunus avium*), evergreen oak (*Quercus ilex*), and English oak (*Quercus robur*), taken when the trees were dormant.

Bare-root trees of beech (*Fagus sylvatica*), considered transplant sensitive, and hawthorn (*Crataegus monogyna*), considered transplant tolerant, were planted into a general tree compost mix in containers. Mulches were then applied to a depth of 4 inches, and results were recorded after one growing season.

Container trials showed that mulch type has a substantial effect on tree survival rate. The survival rate for the unmulched control group was 10 percent, whereas the mulch increased survival rates 20–70 percent in the trial group, with hawthorn giving the best results in all cases (survival, growth, and appearance). The hawthorn trees had 100 percent survival in all cases, and the ones with the hawthorn mulch had 20–30 percent higher dry weights than nonmulched controls. Cherry worked second best, and all mulched trees did better than nonmulched ones.

Field trials of Conference pear and Gala apple trees were mulched to a depth of 4 inches and grown without fertilization or irrigation. Hawthorn and cherry trees again produced the best results, increasing crown volume by 100–150 percent and fruit yields by 400–600 percent. The worst mulch increased crown volume by 20 percent and fruit yields by 50 percent compared to the nonmulched controls.

The breakdown of various mulches releases chemicals that affect plant growth. Other studies have shown that cypress mulch slows the growth of a range of woody plants (e.g., hydrangea, spirea, and viburnum) compared to pine bark.

Eucalyptus grandis mulch has been found to be phototoxic to the germination of a range of seedlings. Black walnut (*Juglans nigra*) inhibits growth and even kills some plants. The tree of heaven (*Ailanthus altissima*) has been found to contain the allopathic chemical ailanthone, which is known as an herbicide.

Some single-species mulches can stimulate the growth of trees. Application of sugars (e.g., dried or liquid molasses) has been shown effective at increasing root vigor and reducing transplant stress and increasing survival rates. Both hawthorn and cherry wood are high in sugars such as sucrose and sorbitol. Also, some pure mulches such as hawthorn have been shown to increase concentrations of enzymes in the roots and leaves that help plants defend themselves from various pathogens.

Other Organic Mulches

Many types of organic materials can be used for mulches, depending on availability and cost. They are all generally beneficial and may be the best type of mulch, depending on the application and/or cost, for a given situation or plant species. Among the other types of organic mulches, the best-quality ones are straw, compost, and newspaper.

Straw—Straw is the dried stalks of grains (which are actually types of grasses) after the seed heads have been harvested. Straw often has a shiny pale gold color and is a good mulch for many purposes. It is generally applied 4–5 inches deep in ornamental beds and 8–10 inches in vegetable beds.

It protects soil, improves soil as it breaks down, and provides good winter protection. Researchers have found that Colorado potato beetles had a much more difficult time finding potato plants mulched with straw as compared to unmulched plants (other research has shown significantly fewer eggs and larvae also). The whitish reflective nature of straw is also beneficial in reducing soil temperature, as it reflects more light. It is reported to protect tomatoes against soil-dwelling diseases.

Hay—A mixture of grasses (and sometimes clover) that is cut, dried, and baled with the seed heads intact (including any weeds present). Hay is often a dull brownish green color. It can be used as a mulch, but the seeds in it often germinate, becoming weeds. Apply 3–4 inches deep in ornamental beds and 5–6 inches in vegetable beds. It will protect the soil, improve soil as it breaks down and provide good winter protection. Many hay fields are now sprayed with an herbicide called picloram, which is sold under the brand names Grazon and Tordon. This herbicide is used to kill broadleaf weeds and persists in the environment on the hay. It is reported that if hay treated with this herbicide is later applied as mulch, the treated hay will still kill many plants (even trees) years after application. To test the hay to see if it is safe to use as mulch, place some in a bucket at least one gallon in size and soak in water for a few hours. Next pour the liquid on any broad-leaved plant and see if they become stressed or die. Peanuts and beans are very sensitive, so they make good indicator plants. When contaminated hay is used on plants, they will have more insect and disease problems even if they are not killed outright.

Newspaper—Works best if shredded first and applied 4–6 inches thick. It is generally free and can protect plants from frosts. Research at the University of Vermont has found a 6-inch layer of shredded newspaper exceptionally good at suppressing weeds for up to nearly two years. Best used as a special-purpose mulch. *Note:* Some inks may still contain toxins and heavy metals. Most inks used on newspapers are now safe and biodegradable; however, unless you know for sure, it's better to be safe.

Compost—Compost is a very high-quality mulch when applied 3–5 inches deep. It has a high nutrient content, improves soil fertility, stimulates plant growth and general health, does not wash out in rain, is weed free if made correctly, offers fair resistance to compaction and excellent resistance to blowing away in wind, contains and stimulates the growth of beneficial soil life (microbes, worms, insects, etc.), and suppresses the growth of many weed species (often better than dangerous chemical herbicides). Research at Ohio State University has found that a 1-inch-thick layer of compost is as effective at disease control as any synthetic chemical on the market. The only downsides are that good compost can be expensive—most gardeners cannot get enough; supply is limited in some areas; and quality varies widely from excellent to very poor. Also, many products are labeled and sold as compost when they are not.

Notes: Very green or partially decomposed (immature) compost is best as a mulch; older, more mature compost is effective to mix into soil layer. Using 1–2 inches of compost directly on soil with 2–3 inches of composted native mulch on top is the best combination possible. Good compost is free of plastic, rocks, trash, and other contaminants.

Compost should have an index of at least 5 on the Solvita Compost Maturity Test for use as a mulch. If used as a soil amendment, compost should have an index of 6 or higher.

Compost, like all other products, can range from extremely good to very bad. For example, Municipal Solid Waste (MSW) compost, while generally very inexpensive, often contains pieces of glass, plastic, metals, and other contaminants. Over time, after repeated applications, these materials tend to rise to the surface and become very unsightly.

Biodegradable Weed Barriers—New biodegradable weed-barrier mulches are entering the market that are made from recycled paper and cardboard. Better ones contain holes for air and water penetration and will last about one growing season. A few examples are:

Hydraulic Mulches: These are a group of special-purpose mulches often used commercially for hydroseeding and vegetation establishment. They are often used in a water-based slurry and mixed with tackifiers (agents that improve adhesion) to help hold seed and fertilizer pellets to the soil for vegetation establishment. They can be applied with spraying equipment to cover large areas quickly and are often used to prevent erosion while vegetation is being established. There are dozens of brands and variations produced by many companies.

Wood Cellulose Fiber: This material is derived from trees and sometimes recycled newspaper or cardboard. Polyester or other synthetic fibers may be added for greater strength. Also used in combination with straw, hay, or other organic mulch. For many years this was the accepted way of stabilizing soil and planting grass to prevent erosion. Several new studies have found that recycled mulches and/or compost work better and at a lower total cost.

Erosion Control Blankets: These are special-purpose mulches made from various types of organic material (fibers) held in place by a mesh (often plastic or polyester fiber). They are used commercially to prevent erosion on steep slopes and disturbed soils. The fiber may be straw, shredded wood, coir, cotton, hemp, or other organic material. These are made into blankets that are shipped in rolls. To apply, the blanket is unrolled at the bottom of a slope and is pinned to the soil. The second roll is applied above the first with a small overlap of the material like shingles on a house. Sometimes the blankets will also contain seed and fertilizer.

Inorganic Mulches

The inorganic class of mulches includes any material that was never alive, such as rocks, gravels, plastic, stepping stones, bricks, pavers, etc. In general, these mulches are best used in special circumstances such as for decoration, pathways, or erosion control. The color of these types of mulches will affect plant growth. Light colors tend to reflect more sunlight (energy), particularly the photosynthetically active or growth-promoting radiation. As a result, air temperatures around the plant are much lower than with darker colors. Dark colors can raise the air temperature around plants 35°F or more, which could make the plant more susceptible to damage from sudden cold weather. The dark colors absorb the energy better, hence soil temperatures are much higher and root growth is decreased.

Plastic Mulches

Inorganic mulches are falling out of favor with experienced landscapers and horticulturalists due to the problems they create and the new research that has demonstrated the benefits of organic and living mulches. Clear or black plastic can be used to warm the soil in spring but should be removed to prevent the growth of harmful fungi and other pathogens in the soil. Methane and other gases produced by the anaerobic conditions can build up, damaging plant roots. The better the soil (more fertile) or the higher the clay content, the greater the problems become with plastic mulches. Inorganic mulches such as gravel may perform well in special situations like rock gardens or certain very dry areas of the country (West Texas, New Mexico, Arizona, etc.).

Inorganic mulches often increase the heat index around plants too much, allow many pathogens to grow better since air flow is reduced, and often can be the cause of root dieback and other fungal diseases. Plastic mulches break down (become brittle) when exposed to the ultraviolet radiation in sunlight. The plastic fragments can create a mess when it starts to break down. Plastic is difficult and expensive to dispose of and generally unsightly. Shallow root systems are often created by plastic, and during drought periods the plants may not survive the stress. Lower crop yields frequently result when plastic is used over other types of mulch, and crops often require more water. Solid films prevent CO_2 from escaping the soil, reducing the benefit from localized concentration to plants. Overall, plastic mulch requires frequent monitoring and the use of a costly drip irrigation system for best performance. Studies have shown that plastic increases soil erosion in areas between plastic-mulched rows. Control of pathogens with this method requires chemical fumigation, which is costly and dangerous. Many reports have found that soil under plastic mulch shows an increase in problem nematode populations. Plastic mulches also kill earthworms by preventing them from reaching the surface to feed. One report stated that melons under plastic mulch require more water (due to higher heat index and larger plants). Since plant parts often stick to the plastic, reuse of the plastic tends to spread plant diseases unless the plastic is cleaned and sterilized. Plastic films tend to break down in the second and subsequent years, allowing cracks to form that weeds can grow through. To be economical, plastic mulches must be applied with specialized tractor-drawn equipment. For effective weed control, it also requires a smoother seedbed than organic mulches, which in turn requires additional labor (i.e., costs). Furthermore, plastic mulch heats the soil too much for some cool-season crops like lettuce, thus reducing yields. If used in early spring, it should be removed after air temperatures warm up. Plastic mulch is most effective in chemical programs with poor abused soil that is chemically and biologically out of balance.

Living Mulches

Living mulches are showing greater and greater promise as we begin to understand natural systems better. Living mulches have many positive environmental aspects that we are beginning to measure and quantify, from reducing erosion to increasing beneficial insects and microbes, water infiltration into the soil, and soil organic matter (humus), and many more. Living mulches are very cost-effective for large areas, as seen in agricultural operations such as orchards and vineyards.

Living mulches are sometimes called cover crops or green manures, and they serve very similar purposes, hence many functions and benefits overlap. In general, living mulches are meant to be mowed and left on top of the soil or left standing, while green manures are to be tilled in.

An appropriate cover crop (or living mulch) planted in late fall will:

- keep the garden green all winter
- prevent erosion
- prevent soil compaction
- control winter weeds
- add large amounts of organic matter to the soil
- reduce surface water pollution (provide a natural filtration system)
- fix large amounts of nitrogen into the soil (certain species only)
- improve soil structure and tilth
- store and recycle nutrients
- increase soil productivity and carrying capacity
- attract many types of beneficial insects that help control pests
- prevent pest species and control certain insect pests
- biofumigate to reduce pathogens

New studies have shown that some living mulches (cover crops like crimson clover) can reduce weed seed germination by 27 percent even after being tilled into the soil. The studies also showed that if the nitrogen supplied by the clover was supplied by ammonium nitrate chemical fertilizer, then weed seed germination increased by 75 percent! Penn State University has been researching living mulches since 1975. New data indicate that living mulches tend to reduce frost damage on many species of plants.

Studies have shown that even a grass cover crop can add over a dry ton of organic matter per year to the soil just from the root mass. In some cases, this can reach over 5 tons per acre per year with another 1–2 tons from the aboveground leaves and stems. Many living mulches can add several times these amounts of organic matter.

For years, many people have sworn by living mulches (cover crops, groundcovers, etc.). Researchers at the USDA in Beltsville, Maryland ("Growth Analysis of Tomatoes in Black Polyethylene and Hairy Vetch Production Systems," 659–663), have done studies comparing hairy vetch (*Vicia villosa*, a winter-hardy legume) and plastic as mulches for tomatoes. The vetch-mulched plots had a longer season and produced up to *twice* as many tomatoes. Vetch is less expensive and more environmentally friendly, and it enriches the soil by adding organic matter and nitrogen. Agricultural Research Service (ARS; the research arm of the USDA) scientists found that for tomatoes, growers using vetch had an average increase in profits of 65 percent in one season compared to those growers using plastic. Other crops that had a strong positive response to vetch mulch were melons, snap beans, peppers, and eggplants. For southern gardens, crimson clover (*Trifolium incarnatum*) is a good option (also attracts several beneficial insects). Additional research at the USDA in Beltsville, Maryland, has shown higher yields and nitrogen levels in tomatoes when a living mulch was used compared to black plastic even when twice the amount of fertilizer was used.

The USDA has found that yields of snap bean raised in mulches from mixed annual winter legumes were comparable to those when synthetic fertilizer was used, and for over three years, the yields were higher than conventional tillage systems. Additionally, the living mulch systems required no water, herbicide, fertilizer, or other treatment until they were mowed. Other benefits from the living mulch system included no runoff or erosion.

Research at Kansas State University has shown that yields from muskmelons were much higher when beef manure was combined with living mulches (hairy vetch, Austrian winter pea, alfalfa, and winter wheat) than when synthetic nitrogen fertilizer was used (*HortScience* 31, no. 1 [February 1996]: 62–64).

Other studies have indicated that the type of living mulch affects the availability of nutrients in successive crops. For example, it has been found that red clover produces twice the amount of available nitrogen to successive crops (such as corn) when compared to oats, rye, oilseed, radish, etc. Yields of successive crops were also increased.

New research is finding that certain plants used as living mulches suppress soil-borne diseases. For example, researchers have found that Sudan grass and sweet corn suppress pathogenic fungi such as verticillium wilt.

Many types of pathogenic nematodes are suppressed by sudan grass, rapeseed, white mustard, Elbon rye, canola, etc. These plants also produce chemicals that are allelopathic, or toxic, to other plants, hence they can also be used for weed control. Using *Tagetes* spp. as living mulch has been found to reduce populations of root-lesion nematodes in a few months to a level where they do not cause significant damage. The effects have been found to persist for several years. It was also found that the seed cost for living mulch was about half that of chemical fumigation.

Studies in the Salinas Valley of California have found that low-cost cover-crop methods can reduce soil nitrate leaching by 37–70 percent in intensive vegetable production systems without hurting yields.

In commercial agriculture, other factors come into play when using living mulches. Tillage, compaction from farm vehicles, and preservation of beneficial insects are all factors, but they are beyond the scope of this book. Many new research studies on these issues have been published as recently as 2005–2010 and can be found in journals at the local library.

Remember that in using living mulches, as in all plants, repeated use of one species in one spot will, over time, increase the chance that certain diseases may develop in the soil. It is a good practice to rotate the living mulch and even use multiple or mixed species at the same time.

Using Living Mulch

Living mulch is often used as a cover crop while the main crop is growing. The cover crop will produce old leaves, stems, spent flowers, seeds, and seedpods as the plants grow. These are often mowed to keep the cover crop from competing with the main crop, add organic matter to the soil, and help build a soil surface mulch layer.

Hay fields, orchards, vineyards, and other types of large plantings are getting good results from living mulches. Some types fix nitrogen for the primary crops; others provide a home for or attract beneficial insects that control pests in addition to enriching the soil.

We need to remember that plants and their root exudates affect the microbes that live in the soil. These effects can often be carried over to the next crop. For example, if the living mulch or previous crop stimulates more beneficial microbes, then disease or even supplemental nitrogen requirements on the next crop can be reduced. Some of the beneficial metabolites produced can result in enhanced growth of the next crop.

Researchers are also learning that cover crops and living mulches can alter the amount and types of pathogens in the soil. Sometimes the cover crop will release antimicrobial volatile compounds that have a *biofumigation* effect. This has been well documented with members of the mustard family (Cruciferae). The gaseous chemicals produced by the plant while it is growing or when it is turned under can kill some types of soil microbes, including many pathogens. Hence, when the next crop is grown, it will experience less disease pressure.

Research is beginning on different aspects of natural biofumigation. Biofumigation uses soil microbes to biodegrade organic material (mulch, root exudates, etc.). Depending on the type of organic matter, some of the breakdown products are volatile gases that adversely affect soil-borne pathogens. Members of the Cruciferae family mentioned above contain compounds called glucosinolates that in the presence of the enzyme myrosinase (which occurs in the tissues of microbes or is produced by microbes) break down into isothiocyanate, nitriles, carbon disulfide, or thiocyanate. Many of these are chemical fumigants with the potential to kill pathogens in the soil.

One study showed that volatile compounds released from soil amended with meadowfoam seed meal completely suppressed sporulation by *Phytophthora ramorum* and *Pythium irregulare*. Soil potting mix amended with only 1 percent meadowfoam seed meal showed striking growth enhancement of conifer seedlings. Another experiment with papaya (*Carica papaya*) and meadowfoam seed meal added at 1 percent by volume to the potting mix greatly stimulated plant growth without suppressing mycorrhizal formation.

Limitations of Living Mulch

Living mulch is not suitable for crops or plants that are short, shallow rooted, or sensitive to low moisture or drought conditions that could be enhanced by the living mulch. Also, if the soils are very sandy or have other types of low fertility, the effects of competition may make living mulches unsuitable.

If used in vegetable production, several studies have found that it is best to delay planting of the living mulch until the primary crop is established. It appears that about one-third of the way through the crop cycle works well for many crops.

Some living mulches have different effects on the soil and thus different benefits. One type may improve water infiltration into the soil better than another. A coarser plant material may reside on the surface longer, since it decomposes at a slower rate. Living mulch with strong, tough stalks (lignin) may encourage fungi in the soil, while a soft grass living mulch may decompose quickly and encourage bacteria in the soil.

Living mulch can be planted similarly to any other plant seed. A loose, friable soil makes a good seedbed. For small areas, after the seeds are spread around, lightly raking the soils can help cover the seed and ensure good soil-seed contact. For large areas, follow standard agricultural practices for your area. Your local agricultural extension offices are often a great source of information.

Some living mulch can be used in winter, but others are best suited for hot weather. Also, it is best to keep the living mulch mowed to prevent seeds from forming, unless the land is going to lie fallow for a while.

Mulch Uses and Applications

Mulches can be applied anytime during the year when trees and shrubs are being planted. For most of the country, the best time would be in mid-spring when the soils have warmed up enough for sufficient root growth. In Houston and along the Gulf Coast, soils stay warm enough for root growth year-round, so the timing does not matter. However, new studies in the South are suggesting that early fall might be the best time so as to trap the soil heat and prevent the soil from cooling off as much during the winter. The fall application of mulch would also help protect the root zone of tender plants better. Some research has indicated that a twice-a-year application at a 2-inch thickness each is better than one application with a 4-inch thickness. Remember that nature mulches in the fall.

A good healthy organic mulch should support fungal growth (white spider web–looking patches, yellow patches, toadstools, mushrooms, etc.) and provide a home for insects, earthworms, and microbes. Research at the Morton Arboretum in Lisle, Illinois, and at Cornell University has confirmed other studies and clearly shown increased soil moisture and root growth under mulches made from tree trimmings (i.e., what has become known as native mulches). Other studies in Illinois and North Carolina in disturbed soils have revealed greater foliage development on recently transplanted trees (sugar maples, crape myrtles, and Callery pears) as compared with unmulched controls.

Research has shown that many plant diseases occur only when a plant is under severe stress (hot soil, lack of moisture, etc.). For example, many common fungal canker diseases of woody plants occur on stressed plants. Repeated studies have also shown that pest insects are attracted to stressed plants. Good-quality and established plants (several years old) are very valuable and are expensive to replace. The traditional cure for insects and diseases is to use toxic pesticides, fungicides, and other dangerous chemicals that are very expensive and require lots of time to apply correctly. Also, who wants a loved child or favorite pet around plants covered with poisons? It is much cheaper, takes less work, and is far safer to just mulch your plants every year and prevent problems from occurring. Properly mulched beds also make a home look much nicer and help add value when it comes time to sell your home.

Mulched beds encourage root growth, conserve moisture, and provide a home for beneficial microbes, particularly in difficult environments. As a result, a good mulch naturally prevents many plant diseases. As better diagnostic tools have become available in recent years, scientists are really beginning to understand how and why mulches work.

Seedbeds

After planting a seedbed, sprinkle a fine layer of mulch (weed-free dry grass clippings, compost, etc.) over the seedbed to help prevent soil from crusting and keep soil moisture more even, thus helping seeds germinate. Large seeds can germinate and grow through a thick layer of mulch, but small seeds cannot. For seedbeds, we need to use mulch that is fine grained and easy for emerging plants to penetrate (i.e., cardboard or newspaper would not be a good choice). Plastic mulches can get too hot and cook the emerging seedlings. A new type of mulch made from recycled paper that is made into pellets is beginning to become available. Since the pellets tend to be loose, they should theoretically work well for larger seeds if the pellets are not applied too thick. However, a problem occurs if these paper pellets are made from recycled paper containing aluminum (most paper currently does), because then they can cause severe reduction in plant growth and seed germination. Some of the fiber-based row covers also work well as a temporary mulch for seed germination.

Paths and Large Areas (Sheet Mulching)

Place sections of newspapers in layers 6–8 pages thick on top of weeds or grass and overlap the layers by at least 25 percent (50% overlap is even better). Opening the folded sections to form larger sheets is far more effective than using the folded sections. It is harder for grass and weeds to grow between the layers to emerge (i.e., it takes a lot more of the energy stored in the stems and roots the farther they must travel horizontally before they can go up). Apply mulch 3–4 inches deep on top of the paper and water in well. The weight of the wet mulch and paper presses down against the grass or weeds and will quickly kill them. This is done first by smothering the plants (little or no oxygen when first applied), second by blocking the light required for photosynthesis, and third by the phytotoxic effects of chemicals in the newspaper. If the mulched area is not thoroughly watered and packed down, some grasses like common Bermuda and nutsedge can live for months. After wetting, it is useful to compact the mulch by walking over the area to press it down tight against the ground. In a few weeks, earthworms and microbes will have devoured the dead grass or weeds, and the paper will have begun to decay, with both types of organic matter enriching the soil. A thick layer of fresh ground native mulch is more effective for killing grass and weeds than composted mulch. Fresh ground mulch requires nitrogen to break it down, hence microbes take the nitrogen from the leaves of green grass or weeds, assisting in the killing of the weeds and grass. The drawback from using fresh ground mulch is that it may contain some weed seeds or acorns that passed through the grinding equipment undamaged and may eventually sprout.

Studies at the University of Georgia have shown that by walking over the same (unmulched) area twenty-seven times, we can compact that living soil to its maximum capacity. Mulches prevent soil compaction by spreading the weight over a much larger area (preventing or reducing the compaction), helping protect the soil and preventing erosion.

Mulches for top-dressing paths and walkways should consist of larger pieces with very little fines (small particles). Fines break down quickly, become mushy when wet, and get stuck on shoes and clothes. Larger pieces are slow to break down (often lasting several years); spread weight, hence preventing soil compaction; lock together better, preventing erosion; and provide hiding places for beneficial insects. Mulches composed of particles that are long and skinny generally work best. Mulches that are flat chips (pine bark nuggets, wood from chippers, etc.) tend to pack together very tightly, preventing the soil from breathing and water from entering.

Weed Control

Many weed seeds require exposure to light before they can germinate. A layer of mulch will prevent light from reaching the weed seeds. Additionally, some earthworm species eat weed seeds, hence a good mulch layer allows worms to find and consume the seeds.

Field studies at Cornell University found that 3 inches of mulch suppressed almost all weed growth. Furthermore, 3 inches of mulch resulted in more shoot growth on transplanted white pine and pin oak saplings when compared with either bare ground or 6–10 inches of mulch, the latter suggesting that more is not always better. A layer of 3–4 inches of settled mulch produced the maximum benefit in their studies. The USDA has found that a 3-inch layer of composted mulch made from green waste (grass, leaves, and trimmings) provided 98 percent control of weeds such as common purslane, Bermuda grass, and redroot pigweed.

Certain types of mulches suppress weeds better than other types, and research is starting to evaluate these weed-suppression properties. Research at the University of Connecticut has found that compost suppresses weeds almost as well as leaves and straw. However, weeds that made it through a compost layer were very healthy. In general, compost that is still "green" (immature, i.e., having a value of 5 or less on the Solvita Compost Maturity Test) will work better for weed control.

Weed control can be enhanced by applying corn gluten meal (a by-product of corn milling) to mulched areas (many nurseries, gardener supply catalogs, etc., now carry corn gluten meal or can order it for you). The corn gluten meal can be applied directly to the soil before the mulch is applied or mixed in with the mulch. Research at Iowa State University has shown that it is a natural herbicide and will prevent or reduce germination of many weed seeds. Additionally, it also has benefits as an organic fertilizer, since it contains 60 percent protein (good for microbes and earthworms) and is 10 percent nitrogen by weight. It works by inhibiting or stopping root formation at the time of seed germination but does not affect mature grasses or plants. Use it at a rate of 15–20 pounds per 1,000 square feet. It needs to be applied in early spring just as weeds start to germinate, as it has a window of effectiveness of about three weeks and it is activated by moisture. The extra nitrogen will help feed the microbes breaking down the mulch, thus enriching the soil. This nitrogen, stored in the body mass of the microbes and in the humus produced from the decay, is slowly released to the plants as they need it.

Research at Auburn University has shown that recycled paper made into pellets or in a crumbled form is also effective for the control of some weeds. As these are new products, there is not a lot of information available; however, the recycled paper in the pelletized form seems to be more effective. As mentioned earlier, part of the reduction in weed growth, particularly on acid soils, may be from toxicity due to aluminum in the paper.

Annuals and Vegetables

For warm-season vegetables like tomatoes, peppers, and eggplants, apply mulch after soil has warmed up (60°F–65°F) in spring. Often in the Deep South (Houston and the Gulf Coast) the problem is keeping the soil cool enough for germination. Applying good thick mulch early in the season keeps the soil cool and moist. Also, different types of organic mulches can assist in disease and pest control for certain vegetables.

After the soil has warmed up in the spring, all vegetables derive benefits from mulching. Using mulch in the vegetable garden conserves water and keeps the soil cooler. It also helps prevent the spread of disease organisms by eliminating the splashing and scattering of pathogens to the leaves of other plants when a raindrop hits bare soil. Although this benefit holds true for all plants, from flowers to shrubs and trees, it has extra value for vegetable gardens, since it is common to have many plants of the same species near each other and susceptible to the same pathogens.

New research has shown that many grasses, annuals, and vegetables do best in a soil dominated by bacteria. Mulches that tend to increase the beneficial bacteria lead to long-term soil health. A few examples of these are green leaves, straw, hay, and grass clippings. However, woody mulches made from tree trimmings (native mulches) are very useful to improve the quality of soils with high clay content and are a useful tool in vegetable gardens.

Research at the Connecticut Agricultural Experiment Station has found that leaf compost can be substituted for synthetic 10–10–10 fertilizers without loss of yields. They also found that composted plots had better yields than those without compost. The beneficial effects of composts continued to increase in the second and third years of application. They also found that when compost was used with inorganic fertilizers, only half as much fertilizer was needed for the same yields (again saving money).

Research has also shown that plants like tomatoes, when grown in an organic mulch like a layer of mown vetch versus black plastic, live longer and develop less disease. These plants have larger, more robust root systems that allow the plant to absorb nutrients better. Other benefits were higher yields, even when using 50 percent less fertilizer.

Experiments at Ohio State University have shown that for tomatoes, organic mulches (i.e., wheat straw, composted bark, shredded newspaper, etc.) had the same weed-suppressing and moisture-conserving abilities as black plastic but cost less, and had several times higher yields than bare ground.

USDA studies have shown that mulched plants were often three times larger than unmulched plants after several years of growth. Other researchers have found that a field mulched with compost had over 1,000 pounds more strawberries than a field treated with the toxic chemical methyl bromide—and at lower cost.

Trees and Shrubs

Research has shown that most trees and shrubs grow better with thick mulch under the dripline instead of grass sod, which is difficult to grow in the shade anyway. Some species of trees, like post oak, are so severely damaged by the presence of turf, it can lead to tree death. Tom Smiley, a plant pathologist, reports that trees are much healthier (fewer disease and insect problems) and experience fewer soil-compaction problems when mulch is used (*Tree Care*

Industry [April 1997]). Previous research reported in *Tree Care Industry* has shown that the microbes found living in native mulches under trees are vital to tree health. Similar results are being found for most shrubs. It is best for most plants if mulch is applied frequently, in thinner layers, until the desired thickness is reached, with composted mulches being the preferred choice. The beneficial mycorrhizae critical for tree and shrub health flourish under leaf compost but grass does not. Numerous research studies have shown the importance of mycorrhizal fungi for increased tree growth and increased resistance to disease. In studies of pine seedlings, it was found that the growth rate increased 700 percent in some species. The native mulches stimulate the growth of these beneficial fungi, while barks sometimes actually suppress their growth.

Mulch around trees has benefits not only for what it provides and keeps in but also for what it keeps out. It is well known that turfgrass and trees do not coexist very well, and we are now beginning to understand why. Healthy turfgrass requires bacterial associations in the soil that dominate the microorganism population at the expense of fungal species, whereas trees and shrubs require soil dominated by fungal species for maximum growth and health. Mulches made from recycled tree trimmings (native mulches) greatly increase the beneficial fungi in the soil required by trees and shrubs. The requirements are exclusive, hence when we try to combine grass and trees, one or both suffer (see Elaine Ingham's work on soil ecology for more detail; soilfoodweb.com).

Research at the Morton Arboretum showed that the root density of trees (littleleaf linden, sugar maple, green ash, and red maple) was greater in a mulched environment than under bare soil and far greater than under turfgrass. Turfgrass inhibits the growth of a tree's fibrous root system, slowing growth, reducing vigor, and making the tree more susceptible to disease, insect pests, and drought. For new freshly planted young trees, a ring of mulch starting a few inches from the trunk and extending out several feet can double or triple its growth rate. Studies have shown that trees and some woody plants can grow up to six times faster if given a 10–20-foot diameter of mulch.

In times of drought or stress, the trees will outcompete turfgrass for water and nutrients in the root zone layers. As most grounds managers know, it is very difficult to maintain high-quality turfgrass beneath the canopy of mature trees. Often the best solution is to separate the species by maintaining a mulch zone out to the dripline of the tree at the outer edge of the canopy. This has another advantage, as it prevents "mower blight" (damage to the tree's bark and cambium from mowing too close).

When applying mulch to a tree or shrub, do not let the mulch come in contact with the plant's bark. When trunks are covered with mulch, the higher moisture conditions created by the mulch often will cause the bark to rot. Unlike roots, bark on trunks was not designed for high-moisture conditions, which allow fungal cankers to grow and attract moisture-loving insects like carpenter ants. Hence, it is best to leave 1–2 inches of open area between the trunks of woody plants and the mulch layer. For larger trees, leaving 6–8 inches around the trees free of mulch will also reduce possible rodent damage. For most trees and shrubs, a mulch layer 2–4 inches deep works well. Avoid hay mulches around trees, as the seeds they contain can attract mice, voles, and other small mammals that may damage the tree bark during their feeding after the seed is gone (also remember that hay mulches stimulate bacterial growth in the soil and trees prefer fungi).

Native mulches are the best for trees, since they are made from tree limbs and leaves and therefore have the exact nutrient makeup that trees need. As the mulch decays, it acts like a perfectly balanced slow-release fertilizer in addition to its other benefits. If the native mulch has been partially composted first, using large piles and long time frames, it will be colonized by many species of fungi that are beneficial to trees, shrubs, and most perennials.

When planting new trees or shrubs, the moderate release of nutrients from a compost-amended backfill material and from the surface mulch will provide all the nutrients the new tree needs for the first couple of years if a good-quality compost and mulch is used. About 2–3 inches of new mulch should be added every year to replenish the mulch that has decomposed (remember, this decomposition is naturally feeding the tree the exact nutrients it requires).

Some experts have found that an organic fertilizer combined with a good mulch can speed the breakdown and enhance the growth in young trees. If the mulch you have available is bark, dyed wood, or is freshly ground, a little fertilizer will help prevent nitrogen tie-up problems. I recommend using organic fertilizers if possible. Seaweed and fish emulsion liquid fertilizers are often a good choice, since they also contain growth hormones and trace elements.

A study at the University of Maryland has found that the insect pests known as azalea lace bugs are attracted to rhododendrons (azaleas) that have received supplemental synthetic nitrogen fertilizer. Many other studies have found that pest insects are attracted to plants that have received synthetic fertilizers. So one of the best (and cheapest) methods of reducing insect problems is to fertilize naturally using compost and mulches. Additionally, for maximum benefit, the mulched area should be expanded by several inches each year as the root zone of the tree expands. After a few years, companion plantings may be added to enhance the appearance in the landscape. The closer we copy the way trees grow in their natural setting (a forest), the better success and fewer problems we have.

Erosion Control and Sediment Fences

Mulches can be very useful for erosion control. Erosion occurs wherever the soil surface is exposed to wind or water. Without some form of cover (mulch or plants) to protect the soil, erosion occurs. Wind removes the finer bits from the soil that contain much of the nutrients and organic matter, leaving behind the coarser particles. The eroded sediments may choke and pollute streams, block stream channels, cover roads, or fill in lakes. Erosion causes damage to the land from which the soil is removed, to the water that transports it, and to the place where it is deposited, hence it needs to be controlled or prevented.

The results of a two-year study released by the Texas Transportation Institute (Texas A&M University) in 1997 found that compost and native mulches were as good or better than all other erosion control methods and much cheaper (Research Report 1352–2F). Similar results were found by the EPA (U.S. Environmental Protection Agency) and FHWA (Federal Highway Administration) when comparing "yard trimmings compost" to hydromulch for erosion control and vegetation establishment. The results showed that compost outperformed the hydromulch with synthetic chemical fertilizer added. Several other studies have found the same results, that compost and native mulch outperform alternative methods and at lower cost.

Mulches that are woody (i.e., ground-up tree trimmings), with long thin pieces and frayed ends, tend to work best for erosion control. These characteristics allow the mulch to physically lock or mat together. Additionally, fungal fibers will grow, which also lock mulch particles both to each other and to the ground after exposure to water.

Freshly ground native mulch placed in a small pile or in long rows along a construction site is very effective at removing sediment from storm water runoff. When runoff carrying sediment (often valuable topsoil) hits the mulch pile, its velocity is slowed. As a result, the water no longer has the energy to carry the sediment, and it is dropped inside the mulch line. Also, as the water passes through the mulch, the filtering action removes additional sediment and dissolved minerals (nutrients). Several studies have shown that mulch is more effective than sediment fencing and often costs less. When the construction is finished, the sediment (topsoil) that has been enriched by the decomposition of the mulch into compost can be spread on the landscape or used in the construction of flowerbeds and other plantings. This also results in better-quality surface water in our rivers and lakes, as fewer contaminants and sediments flow into them.

Developers and builders are also finding that coarse-ground mulch made from ground-up recycled construction wood makes good filter berms and work areas. When covering a work area, it prevents mud from being tracked into the building under construction. By the time the building is completed, it has decomposed enough that there is no need for removal, as it just adds organic matter into the soil, which helps with the landscaping and turf establishment.

Soil Development and Health

Scientists at Laval University in Quebec, Canada, have found that when "ramial wood chips" (RCW; derived from a French term meaning "twig wood"), made from small branches of deciduous trees that have been ground up, is applied to the soil, their lignin is converted rapidly by certain soil fungi into valuable humus ("The Use of R.C.W. in Agriculture," 1–10). Ramial wood is essentially native mulch made without large limbs and logs. Experiments for twenty years in Africa, Canada, and Europe have shown that a 1-inch layer of chopped or crushed twigs mixed into the top 2 inches of the soil will be broken down by aerobic fungi called basidiomycetes. Benefits found include improved soil aggregation and moisture retention, larger root systems, more mycorrhizal associations with roots, decreases in some soil pathogens, and yield increases from 300 percent (strawberries) to 1,000 percent (tomatoes) compared with untreated soil. Garden prunings and twigs should be chopped or ground up, but the leaves should be removed (the breakdown of green leaves encourages bacteria that can displace the basidiomycete fungi). This type of mulch is best applied in the fall so the soil is ready for use by spring (temporary nitrogen tie-up may occur when first applied). This approach works well in all but wet soils. Branches from conifers (pines, etc.) should not be used, since their lignin breaks down into polyphenols rather than beneficial humic and fulmic acids (this also applies to most bark products like pine and hardwood). Native mulch produces the same benefits, as it is composed mainly of the twigs and small limbs from primarily deciduous species. Other researchers have found that simply applying the mulch on top of the ground works well—with less effort.

Research into soil ecology has also found that the basidiomycete fungi produce their sexual spores on a large fruiting body, or basidium. These fruiting bodies are characteristic features of many healthy forest soils and are found in meadows in the spring and fall. Many of these Basidiomycota form ectomycorrhizal associations (beneficial symbiotic relationships) with tree roots and are *critical* to the nutrition, growth, and health of the tree. These extremely beneficial fungi only grow on native (ramial) mulches. It is this group of fungi that convert lignin (white rot) and cellulose (brown rot) into valuable humic acid, fulmic acid, and humin that we call humus. If we apply fungicides to our trees, plants, or grass, we end up killing these valuable friends (sort of like getting ready to run a marathon and taking a pistol and shooting ourselves in the foot first).

Clay soils have been found to rapidly improve in all aspects of soil quality and health when native mulch is applied. Sandy soils respond well to compost mixed into the sand and compost used as mulch, with a top dressing of 1 inch of native mulch. If the native mulch has been composted for a few months, it will work even faster and provide additional benefits.

Many soils develop a condition called hardpan after repeated exposure to excessive use of synthetic chemicals. The soil becomes extremely tight and often very hard, hence the name. Air and water cannot penetrate the soil, and beneficial microbes and animals are limited. This hard layer can occur at the surface or inches below the surface.

In general, all types of organic mulches will improve all types of soils. Some just work faster than others. The amount of mulches used, how they are applied and handled, and the starting condition of the soil are all factors in soil improvement. As these materials break down, they eventually become soil organic matter that we call humus. For healthy soils, the organic matter should be broken down enough that there are about twenty-five to thirty carbon atoms for every atom of nitrogen present. We call this the carbon-nitrogen ratio, expressed as C:N (i.e., 30:1).

As organic materials age, they break down into different types of chemicals that enter the soil. These constituents are grouped into some basic classes: cellulose; hemicellulose; lignin; water-soluble fractions such as simple sugars, amino acids, and aliphatic acids (succinate and acetate); a protein fraction; and an ether- and alcohol-soluble fraction (fats, oils, waxes, and resins). As organic material gets older, the content of the first three fractions increases and of the latter groups decreases.

The native mulch (if composted first) makes an ideal ingredient to add to prepared soil planting mixes in small amounts. It helps lighten the soil mix, improves aeration and looseness, and, most importantly, provides a long-term energy source (carbon) to help promote a healthy soil food web.

Erosion and Compaction

Rain drops can hit the ground with velocities between 10 and 20 mph (miles per hour) in normal conditions and over 60 mph during severe storms. The kinetic energy associated with the raindrop increases as the square of the velocity (V^2). If the soil is unprotected by a mulch layer, the impact dislodges the soil and erosion begins. Compare this to the surface runoff on near-level areas where the rainwater only moves about 1–2 mph. Of course, on slopes and in gullies and streams, the water can move much faster and carry away the valuable topsoil knocked loose by the raindrop.

Another effect associated with raindrops hitting bare soil is surface sealing. This occurs when the dislodged soil particles wash down into the soil pore space and clog it up. This creates a thin compacted zone at the surface that seals off the soil, preventing rainwater from soaking in, which greatly increases runoff and causes more erosion.

When this seal dries, it can become very hard and create a crust that can prevent seeds from germinating and penetrating the layer. Since water cannot easily enter the soil, many roots and microbes can suffer or die from water stress or lack of water. Air flow into the soil is reduced, preventing oxygen from reaching plant roots and microbes. It also allows gases to build up that are toxic to many plants and soil animals. The resulting conditions favor the growth of pathogens in the soil.

Decomposition of Mulch

As was mentioned above, there are many beneficial microbes at work breaking down mulch and organic matter into soil components. Inexperienced gardeners often say, "The mulch was not any good; it just rotted away in only one year." Mulch and plant residues (litter, leaves, twigs, branches, root detritus, and exudates, etc.) provide carbon (the energy source), which is the fuel for the soil food web that cycles and stores nutrients, creates soil structure, and prevents pathogens and pests from taking over. When we burn wood logs in our fireplace, the carbon in the wood is combined with oxygen in the air, releasing energy. The same thing happens in the soil. Carbon from decaying organic matter is combined in the bodies of microbes with oxygen from the air, giving them the energy needed to create soil structure and fight pathogens and pests. If we do not feed our army of beneficial microbes, then they will die and the pathogens and pests will take over. In other words, "WE *WANT* MULCH TO DECAY!" In a healthy soil with a good-quality mulch, about two-thirds, or 50–60 percent, depending on the climate, should decompose in a one-year time frame. The remaining mulch will break down at a much slower rate, providing other long-term benefits to the soil. If the mulch does not break down, then the soil is very unhealthy, or you purchased very low-quality mulch.

The rate at which mulch or any organic material breaks down is dependent on many factors: the type of material, the material's age, the particle size, the nutrition content such as nitrogen, the soil moisture and temperature, aeration, pH, and a few others.

Vertical Mulching

Vertical mulching is the name given to a group of related techniques used to inoculate soils and plants with beneficial microbes, like mycorrhizal fungi, and to get organic matter deep into the soil and increase aeration. Numerous and repeated studies have shown that many species of microbes are essential for plant health and growth, insect and disease resistance, drought tolerance, etc. Soils that have been exposed to cultivation, pesticides, fungicides, herbicides, or synthetic chemical fertilizers, or that have low organic matter content, are frequently deficient in these extremely valuable microbes.

Vertical mulching is achieved by drilling a narrow hole (or taking a soil core) 1–3 inches in diameter and 12–24 inches deep. Often compost (or com posted woody mulch) is mixed with soil and a substrate containing the microbes (a water-holding gel, for example). This mixture is placed into the holes, hence the name vertical mulching. Typically these holes are dug 1–2 feet apart around a plant or tree out to at least the dripline (even farther is better). The compost and mulch provide food (organic matter) and energy (carbon) for the microbes as they become established with the plants roots. Sometimes an organic fertilizer and trace minerals are added to aid in deep root feeding.

A variation of this technique can easily be done to improve poor and compacted soils and break up hardpan. First we take a garden hose or, better, a hose fitted with a ½-inch inside diameter steel pipe to make a water drill. We turn on the water to a fast setting and press the tip of the steel pipe to the ground. The water pressure coming from the end of the pipe will bore a hole into the ground, so we just keep turning the pipe and pushing down till we reach our desired depth (in most cases, 12–18 inches deep is sufficient); the hole has to be deep enough to reach the moist soil layer. The water should have created a hole 1–2 inches wide. We move the pipe over 12–18 inches and repeat the process until we have covered all the area we wish to treat. The holes can now be filled in with compost and then watered in. Using a seaweed-based liquid organic fertilizer and some molasses mixed in with the water will jump-start the microbes to improve the soil. A secondary benefit of using a water drill is that the water washes away the soil from the roots without destroying them, whereas a conventional drill or soil core cuts through the roots, creating stress for the plant or tree. *Note:* If the bottom of the pipe has small notches or teeth, it will chew through compacted soil easier. Notches can easily be added with a hacksaw.

Another variation of this technique can be used to lift water from a shallow water table or layer into the root zone, where it can be reached by plant roots. Tests have shown that a mixture of rice hull ash and good compost can have a strong wicking action and actually lift water several feet from the subsoil into the root zone. This type of mulching can be extremely valuable in areas that are hard to water or during drought conditions.

Potting Media

For years, professional growers used pine (conifer) bark as a soilless medium for growing plants in pots. When the pine bark was many years old and well rotted, it worked very well and was inexpensive. Over the years, as the old stockpiles of pine bark left over from the lumber and paper industry have been used up, suppliers have been forced to use fresher and fresher bark. As a result, growers in many areas have been paying a lot more or have been receiving lower- and lower-quality materials, resulting in higher costs and more disease and pest problems. Eventually these higher costs are passed on to the consumer, *and* we have increased pollution from all the chemicals that are required.

With the cost of oil rising, many lumberyards and paper companies are using the bark as fuel for their boilers and dryers. As a result, they are no longer selling bark for potting media, forcing many growers to look at alternative media. To address these issues for growers, many universities have been experimenting with compost and composted wood chips or mulch as a planting and potting medium. It has been repeatedly found that these alternate products work better and at lower cost in most cases. To be successful, growers need to reevaluate their watering and fertilization systems and how they use materials. There have been so many success stories from research projects by various universities. *American Nurseryman* magazine in recent years has published several articles encouraging growers to experiment with compost and composted mulches as a growing medium.

For most gardeners, equal parts of compost, sand, and topsoil will make a good starting mix. Additionally, well-rotted or aged (composted) native mulch works well in many cases.

Jason McKenzie of the Pineywoods Nursery near Conroe, Texas, has found that a 50:50 potting mix of fine-screened composted native mulch and fine-screened compost gives superior growth and root development on many species, with minimal fertilization and greatly reduced watering.

Researchers at the University of Vermont have shown that it is possible for disease transmission to occur via wood chips taken from infected trees and used around healthy landscape plants. They found that the nematode that causes pine wilt (*Bursaphelenchus xylophilus*) could move from infected chips to young Scotch pines if the infected chips were tilled into the soil during transplanting or applied against a trunk that had been wounded. While it is theoretically possible, it is very unlikely that this type of disease transmission would occur in practice, as raw wood chips are not incorporated into the soil, and the mulch should not be piled against a tree trunk (nematodes cannot move more than a couple of centimeters on their own). Additionally, the pine sawyer beetles that transport this nematode are not attracted to wood chips. This small risk can be eliminated by composting the wood chips for a few weeks before applying them. It should be noted that the same risks apply to bark mulches, since infected trees are often the first ones harvested for lumber and pulp.

Verticillium wilt is a common disease caused by a soil-borne fungus that results in the decline or death of many shrubs and trees. It is caused by the fungus *Verticillium dahliae* and endures in the soil in infected plants or as flecks of scleroti, a type of fungal tissue designed for long-term survival of the fungus. If diseased trees are ground up and used for mulch, it is possible that this disease can spread to mulched plants. It has been found that excess synthetic nitrogen fertilizer favors development of this disease. However, this disease is rapidly destroyed if the mulch has been composted for at least three days at a minimum of 130°F.

Rhizoctonia solani is another plant pathogen that causes damping-off of many types of seedlings. This pathogen is actually stimulated by fresh mulches, as it feeds off the cellulose in the wood. Again, composting the mulch for a period of time before using it eliminates this potential problem. Prevention of disease is another good reason to use composted mulches wherever possible, if you can find them.

Remember that many types of mulch develop various types of fungus on their surface as they decay. A few common types are artillery fungus, bird's nest fungus, slime molds, puff balls, toadstools, and mushrooms, among others. Visible signs of fungus are often the fruiting spores and are beneficial to the soil and plant health.

For example, fungi known as the stinkhorn fungus (*Phallus impudicus*) are often found growing on mulches with a high carbon-to-nitrogen ratio. It may start as an egg-shaped mass and turn into a stalk topped with a slime-coated head. This fungus deserves its name, as it often has a strong odor similar to that of rotten meat. The fungi produce this strong odor to attract flies and other insects. As the insects crawl on the slime, they pick up fungal spores that they then carry and spread to other locations. This fungus is most often found during warm, moist conditions in the summer and is actually hard at work breaking down the organic matter in the mulch into a form that plants and other microbes can use.

Fungal activity is the sign that nature is hard at work releasing the nutrients and energy stored in the mulch, which is required for good plant health. If the appearance of the fungus bothers you, the visible appearance can often be eliminated by raking the mulch layer, blasting it with water from the garden hose, or both. Also, if there is a lot of raw or fresh wood (high C:N ratio) in the mulch, the fungus can form a hard barrier that is difficult for water and air to penetrate.

In rare cases, large amounts of organic matter may actually increase disease rather than suppress it. The process by which this occurs is not fully understood. The soil environment is changed by the organic material, and then a rapid growth of the microbial population occurs, using up all the available oxygen (O) and producing large amounts of carbon dioxide (CO_2) in the process. This happens when the material is compressed or so saturated with water that air movement is restricted. Under these conditions, disease organisms would have an advantage for a while.

Pitfalls of Mulching

As the popularity and benefits of mulch have become better known, mulched landscapes have become very common. However, using mulch is a science, not an automatic guarantee of successful gardening. Misuse of mulch, ranging from improper choices to misapplication, may lead to problems. Problems using mulches may occur when recommended horticultural practices and procedures are not followed.

Odors

Odors are warning signs of low-quality and potentially dangerous mulches and composts. Pay attention to the following kinds of odors and what they indicate.

1) Anaerobic organic acids that have a strong odor from putrefying organic matter. The odor varies, depending on feedstock or material and what is going on; however, they are all very bad. These types of

organic acids form under conditions without oxygen (fermentation) that also produce alcohols. Plant roots are very sensitive to alcohols, for as little as 1 ppm will kill most plant roots.

Acetic acid	Vinegar smell—loss of N_2 and P; alcohols present
Butyric acid	Sour milk smell—alcohols present
Valeric acid	Vomit smell—alcohols present
Putrescine	Rotting meat smell—alcohols present

2) Ammonia—implies an immature compost (phytotoxic) and a loss of nitrogen
3) Rotten egg (H_2S)—implies an immature compost (phytotoxic) and a loss of sulfur

Color

Color is often an indicator of potential problems with mulch or compost and other organic materials. A black color does not occur naturally in mulches or compost under good conditions, only a deep chocolate brown. However, many people believe black is good, and some unscrupulous vendors like to take advantage of this idea.

Black organic materials occur in nature when materials decompose under anaerobic conditions (without oxygen). These conditions favor disease and other pathogens and use a different set of microbes to decompose the material. As a result, pure "black" compost or mulch does not have good fertility and indicates anaerobic decomposition as well as other problems. The sulfur is outgassed as H_2S, nitrogen is gone (NH_3) or in the wrong form, and alcohols are usually present. Good compost is a deep chocolate brown when dry.

Industrial wastes are often used to blacken products for marketing purposes (e.g., even though it is illegal in some states, some companies grind up old railroad ties to help darken material). Smelter wastes are sometimes used as feedstock to blacken products. Copper sulfate ($CuSO_4$) or other sulfur compounds may be present. As they break down, elemental sulfur (S) may be produced, which is a natural fungicide that kills the beneficial fungus.

Boiler ash (bottom ash) is another industrial waste product used to color or blacken products. Boiler ash tends to be high in salts and extremely alkaline. The alkalinity is so strong that it will chemically burn raw wood black in a couple of days. The products that contain this tend to be alkaline with high salt and very high carbon-to-nitrogen ratios. Some ashes may contain large amounts of heavy metals that contaminate the mulch, exceeding federal regulatory levels for safety. These mulch products will often turn a bleached grayish color in a few weeks after exposure to sunlight. These types of products are very common in many areas.

A question we often hear is, "What really happens if I use cheap or bad mulch?" Another is, "I do not see any difference, so why should I pay more?" or "I can get it cheaper down the street at your competitor." At the same time, we often hear these same people complain about all the weeds they have and the excessive time they spend weeding, or how much money and time they spent at the doctor's office for an illness or allergic reaction related to the herbicides they used to spray the weeds or other chemicals that may be present.

The following guidelines will help to ensure success:

- Mulch choices and practices vary, depending on many factors like climate, plant species, age of plants, soil type, location, watering practices, and others. These factors vary regionally (state to state) but can also vary in a single backyard.
- Heavy mulch around certain plant species during extreme wet conditions can hold too much moisture. This adverse condition typically occurs in plants adapted to dry conditions (cactus, mesquite, etc.).
- If mulch is applied against the bark at the base of some woody plants, it can lead to stem rot. Some people recommend that for hostas. Mulch should be applied right up to the crowns but not over them or touching them (allow a small gap for air to circulate). Remember that in a forest, the leaf and twig litter (i.e., mulch) is thick under the leaf canopy but becomes very thin at the base of the trunk, hence this is how mulch is to be applied (in other words, copy nature).
- Covering the soil with mulch too early in the season with certain vegetable species can hold them back by keeping the soil too cool (it helps other species grow faster and produce more). When using any soil amendment (compost, mulch, fertilizer, etc.), one must understand the cultural requirement of the species of plant being grown to get the best results.
- If some types of mulches are applied deeper than 4 inches, feeder roots often grow into the mulch layer. Later, a disturbance of the mulch or drying out of coarse mulch layers may injure or kill these feeder roots.

Black polyethylene roll-type plastic mulches often look bad and absorb excessive heat (if not covered by an organic mulch), essentially cooking the root systems of most plants in hot climates. In wet years, the plastic often traps too much moisture in the root zone, drowning plant roots and creating a breeding ground for disease. The perforated types often only work as a weed block if installed a certain way, and most need a large overlap of material to prevent roots/weeds from growing between layers. The use of plastic mulches creates indirect and hidden costs to society related to environmental issues, such as the direct cost of removal, collection, and waste disposal. Recent studies are finding that while plastic mulches help to obtain yields earlier in the season than bare ground, total yields over the entire season are often higher from bare ground.

Always go look at mulch before you order. Sometimes people have a visual image in their mind about what they want, and very frequently they use terminology incorrectly. Also, in different areas of the country, words and terminology are used differently. As a result, consumers will place an order, only to be disappointed in the results when it is delivered.

Furthermore, many dealers and producers will use incorrect or misleading terminology. Some suppliers/dirt yards sell products that use words like *black* and *humus* in their names. These products are often made from fresh pine bark fines, do not contain any humus, and are chemically burned to turn it black by adding very alkaline chemicals (i.e., it is mixed with boiler ash, which is very alkaline and contains high levels of salts). Other dealers will grind up old pallets, scrap wood, trees, and so forth and mix it with fly ash or bottom ash, then sell it as a black hardwood mulch. These kinds of products are very poor mulch choices and are often toxic to many plants. People use them, but when the plants get sick and die, they think "I just do not have a green thumb." People buy them because they are often sold at bargain prices, but they are not very cost-effective.

Mulches made from almost any type of old pallets, scrap wood, tree trunks, etc., will contain very little nitrogen and thus even less is available for the plants. These types of materials often have carbon-to-nitrogen ratios of 500:1! For microorganisms to break down these types of products, they must use up all the available nitrogen in the soil, leaving the plants very nitrogen deficient and stressed. This condition causes the plants to become much more susceptible to insects and diseases. For comparison, good mature compost will have a carbon-to-nitrogen ratio of about 25:1, one that plants love. There is also a risk factor that mulches made from old pallets may be contaminated with dangerous chemicals that are toxic to plants and the environment.

Mulch made from fresh ground trees can also cause problems in some cases. In fresh wood, there is an abundance of soluble carbon compounds that can accumulate in the soil as the fresh wood chips break down. Many beneficial microorganisms only antagonize soil pathogens when they are stressed (i.e., must work for their food). These types of carbon compounds are like candy to the microbes, and while the microbes are busy eating, they do not have time to bother with pathogens. While the good microbes are busy, the pathogens can build up in the soil and are allowed to gain a foothold. Fresh mulches can also become slimy, hold too much moisture, and block airflow, creating conditions for disease organisms to grow. Several universities have found that these effects are worse on soil low in organic matter, new landscapes, and compacted soil. This is another reason why composted mulches are more effective and a better value than fresh mulches.

We sometimes see mulches advertised with phrases like "decay resistant," "does not attract insects," "will not grow mushrooms," and so on. Translated, this means those mulches have been treated with toxic synthetic chemicals (herbicides, pesticides, fungicides, etc.) to produce these effects. These types of mulches defeat many of the benefits that mulches provide, and the chemicals they contain pollute the environment and endanger human health.

Research from the USDA has found that if hairy vetch was killed with glyphosate (the active ingredient in Roundup) and then cut and used as a mulch, yields in some plants were reduced by 50 percent ("Snap Bean Production in Conventional Tillage and in No-Till Hairy Vetch," 1191–1193). Research at Michigan State University, published in the same issue of *HortScience* ("Effect of Pesticide-Treated Grass Clippings as a Mulch on Ornamental Plants," 1216–1219), has found significant growth reduction on ornamental plants using pesticide-treated grass clippings as a mulch, with post-emergent herbicides causing the most damage. It has been learned that most herbicides do not break down as previously believed; hence, we should not use mulch made from plants or grass treated with herbicides (or pesticides or fungicides). Research at North Carolina State University has shown that herbicide-treated grass, when used as mulch, substantially reduces plant growth (80% for cucumbers, 65% for marigolds, 34% for salvias). Purdue University has expanded these studies and found that even growth regulators will persist for months and cause harm when plants or leaves treated with them are used as mulch (*Journal of Environmental Horticulture* 15, no. 4 [December 1997]).

Some media articles have talked about the danger of using organic materials in agriculture and horticulture. Though not a pitfall of mulches or compost, this danger is related to poor management practices in raising chickens and other livestock. Certain new types of diseases may be present in some types of organic matter and particularly in manures. A new strain of *Escherichia coli*, a bacterium that is normally benign or beneficial, has been discovered that is extremely toxic. It has already caused severe illness and death across the country. It is known as *E. coli* O157:H7. As for all *E. coli* strains, it is easily destroyed by heat. This is another advantage of using any heat-composted mulches (e.g., composted native), as *E. coli* is destroyed in a hot compost pile.

A problem that sometimes occurs in mulch is called the "toxic mulch syndrome" or "sour mulch." This most often occurs with bark mulches (pine, hardwood, etc.) but can happen with almost any organic mulch. It occurs when a fine-grained (small-particle-sized) mulch is stacked over 6 feet high and remains wet for long periods of time. The material compresses and starts fermenting (anaerobic decay instead of beneficial aerobic decay) and produces chemicals (methanol, acetic acid, ammonia, hydrogen sulfide, and others) that can kill annuals and damage many woody perennials. These chemicals have a strong sour acid odor in contrast to the pleasant musky smell of fresh-cut wood or compost. This problem can also occur in bagged products that have been stored in wet conditions as well as in bulk products. Using coarse-ground materials reduces this risk and is better for most plants. Bedding plants and low-growing shrubs are the most vulnerable to poisoning by soured mulch, with symptoms appearing a few hours to several days after mulch is applied. Therefore, avoid mulch with a strong odor of vinegar or rotten eggs. Good mulch will smell like freshly cut wood or have a rich earthy smell if it has been composted.

If mulch is piled too deep or if its texture is too fine (i.e., ground pine bark or sawdust), it will easily become compacted, preventing the soil from breathing. When this occurs, the soil becomes oxygen depleted, causing roots and beneficial microorganisms to die. This leads to increased plant stress, which we sometimes see as insect and disease problems. Studies at Cornell University have found that soil oxygen depletion under wood chips (not bark) is minimal, even when piled as deep as 10–18 inches. *Note:* Aeration is a function of particle size; coarse-ground mulches will breathe better than fine-ground mulches and will also allow water to soak into the soil more easily.

During the decay process, various types of fungus may grow on the mulch surface. The artillery fungus (*Sphaerobolus stellatus*) is also known as the shotgun fungus because it can blast its spores 10–15 feet into the air. These spores are brown to black and very sticky, hence they can discolor light-colored surfaces by sticking to them (bird's nest fungus will also shoot its spores but not as far). If discoloration does occur, a soap-and-water solution will help to loosen the fungal spores so they can be scrubbed off. Most visible signs of fungus will naturally disappear as the mulch continues to decay into humus. The appearance of slime molds is distasteful to some people, but the visible signs of this fungus are easily removed by periodically raking the mulch.

These types of problems are much more common on mulch made from fresh or woody material rather than composted. They also are more common in thicker mulch layers (4–6 inches deep).

When using mulches, we need to remember the area of the country we are in (i.e., weather and climate) and the type of soils and plants that are growing. Along the Gulf Coast, a 3-inch-thick mulch may be desired for most species, but in Arizona or New Mexico, an organic mulch 3 inches thick may prevent the scant rainfall from reaching the soil or trap moisture around plants used to very dry soils (e.g., cactus), thus increasing the possibility of disease. In drier areas, perhaps only a 1–2-inch mulch layer would work better, depending on the plant species and watering requirements. Another potential problem in very dry areas is the possibility of fire from a dropped cigarette or sparks from a fireplace or barbeque pit. Mulches like bark, shredded wood products, straw, pine needles, ground rubber, and some plastics are easily ignited.

Another factor to remember is that hot, moist climates have a much faster rate of organic matter decomposition and require more frequent mulching. In cooler climates with long winters, the mulch will break down at a much slower rate. If you are not sure, you can contact the local county extension agent's office or the horticultural department at a nearby university for detailed advice for a given locality.

A few mulches have been shown to hurt plant growth (allelopathic). These include those made of black walnut, eucalyptus (blue gum), tansy, wormwood, and French marigolds.

Playground Mulches

Many types of material are used to help reduce injury at playgrounds. These materials range from sand to gravel and wood chips. In tests done and published in the *Handbook for Public Playground Safety*, it was found that wood mulches perform well. A 6-inch layer of uncompressed wood mulch had the shock-absorbing capacity to protect from a fall of 7 feet, while sand or medium gravel would only protect from a 5-foot fall. Uncompressed wood mulch 9 inches deep would give the same protection from a 10-foot fall. When using any material, it is best to use a little extra to give an added margin of safety. Also, wood mulches tend to be a little cleaner than sand or gravel. For this type of application, mulches made from fresh ground materials work best, since they take the longest to break down. Mulch for playgrounds should have the fine particles removed and the very large particles limited.

Using Mulch for Frost Protection

Along the Gulf Coast, there are only a couple of nights a year when frosts are a problem. As a result, many gardeners are tempted to grow tropical and other tender plants like citrus. There is an easy trick that I learned from a friend that can give gardeners a few extra degrees of protection on cold nights if the plants have been well mulched.

It is well known that a thick mulch layer will keep the soil warmer in the winter, since the mulch acts like a blanket (an insulator) and keeps the heat stored in the soil from escaping. On the night of the cold weather, take a garden rake and remove the mulch out to the dripline of the plant, exposing the soil underneath the plant. This will allow heat energy stored in the soil to escape by both radiation and convection. *If* there is no wind (or it is very light) and only a couple of degrees of protection are needed, there will be a chimney effect created around the plant from the rising warmer air. The energy that is released from the soil will keep the air around the plant a few degrees warmer, and often it is enough to prevent frost from forming. If the soil has been well watered a few days before it turns cold, then even more energy will be stored and be available to help warm the plant. The day after the cold weather has passed, rake the mulch back into position around the plant. This allows the heat energy from the surrounding soil to move into the area around the plant and recharge the soil for the next time it is needed.

A variation of this technique is to remove the mulch as before, creating a circular berm around the plant with the mulch forming a basin. If the soil type is slow to drain, this basin can be filled with water. The heat energy stored in the water may also give some extra protection.

Another benefit of using mulch is protecting soil structure. The occasional freezing and thawing of soil under mulch, if done slowly, can be beneficial to the soil. However, under the conditions of rapid freezing and thawing, the soil aggregates can be broken down, reducing the tilth of the soil.

Mulches for Animals

Various mulches are often used for animals, from bedding materials in stables to flooring in dog pens. Since animals are sensitive to things humans are not, it is best to use natural or organic mulches.

Traditionally, mulches such as pine shavings have been used as animal bedding. However, people are finding that fresh ground native mulches that have been screened to remove the larger pieces work better than wood shavings. Fresh ground native mulches help absorb urine and feces from the animals, and the microbes that live in them will help break down the manure and urine, resulting in less odor. As a result, the mulch lasts much longer, saving time and money, as one does not have to replace it as often. When the pens are cleaned, the spent material is great for composting. This not only saves disposal costs but creates a useful by-product. Dogs and cats also like to chew on native mulches to help clean their teeth, which helps reduce their chewing on living plants.

Another use of screened native mulches is on horse arenas. Several inches of mulch last a long time, have the benefits listed above, and are much easier on the hooves of the animals, thereby reducing problems and veterinary costs.

Fine-screened native mulch also works well as bedding for hamsters, chickens, rabbits, and other smaller animals.

Many people have found that some mulches work well for animal control, as they help keep them out of flowerbeds. The most common mulch used for this purpose is cracked nut shells, such as from almonds, pecans, hickory, walnuts, etc. The hard, rough edges hurt the animals feet, so they stay out of the treated area. It only requires a light sprinkling over the regular mulch. Another mulch that has been used is ground-up fresh green pinecones, as the sharp barbs also hurt their feet. For severe problems, one may lay down chicken wire before sprinkling the nut shells, then the animal's claws also get tangled, which they do not like.

Most organic mulches are safe for animals to eat, but some may be toxic (e.g., cocoa shells to dogs).

Mulch for Bioremediation

Mulches contain lots of carbon, and when carbon is combined with oxygen from the air, energy is released. It is this energy that allows microbes to break down many types of chemicals and clean the air, soil, or water.

As the supply of freshwater becomes scarcer and more valuable, more and more communities are turning to gray water for landscape use. Gray water is all the wastewater from our homes and buildings *except* that used for toilets or sewage disposal. It often contains soap residues and other cleaners, food particles, water softener chemicals, and much more. There are two ways to clean the water for reuse: (1) buy an expensive treatment and purification system coupled with a wetlands-type filtration system or (2) just use mulch.

Mulch can be used in two ways. The first is to form a heap (as in a compost pile) and let the gray water filter through it, then we can collect and use the filtered water. The other way is to let the gray water filter through a thick mulch/compost layer directly into the flower beds. The best way to mulch flower beds is to apply a couple of inches of compost on the soil followed by a couple of inches of woodier mulch on top of the compost. This is also ideal for treating gray water.

As the gray water filters down through the mulch layer, the tremendous number of microbes (fungi and bacteria) that live in this layer eat the chemicals in the gray water, breaking them down into harmless components. The microbes also help buffer the soil from changes in pH due to the effects of soaps and other cleaning products that tend to be alkaline. By the time the water reaches the plants' root zone, it is much safer for the plants to use.

Studies have shown that a well-drained loamy soil can absorb 500 gallons of gray water per week for every 1,000 square feet of flower bed. Since many households produce over 100 gallons per day, this is enough water to meet the demands of most plants. If more gray water is produced than can be used, it can be filtered through a compost pile and then used to water the lawn or released to lakes or streams.

Gray water should be treated and used as it is produced so that pathogens do not have time to grow. If it must be stored before use, then some form of treatment should be considered to prevent bacteria and pathogens from growing.

Many types of chemicals can be broken down by microbes that live in compost and mulch. Oil and grease will be absorbed by the mulch and are easily broken down. It has been proven that compost can bioremediate (in situ or at a facility) the following types of chemicals:

- Polynuclear aromatic hydrocarbons found in creosote
- Carbofuran pesticide (carbamate family) and simazine herbicide (triazine family)
- Polychlorinated biphenyls (PCBs); trichlorphenol (TCP); and Benzo[a]pyrene (BaP), a polycyclic aromatic hydrocarbon (PAH)
- Pentachlorophenol (PCP; a wood preservative more toxic than chromated copper arsenate, or CCA), dioxins, cyanides, DDT, TNT (explosive), creosote, and coal tar
- Explosive propellants (WC860 and H5010) that contain nitrocellulose, nitroglycerin, dibutylphtalate, calcium carbonate, dinitrotoluene, diphenylamine, potassium nitrate, sodium sulfate, graphite, tin dioxide
- Hexachlorocyclohexane (HCH)
- Alachlor, metolachlor, and 2,4-D
- TCE (trichloroethylene, solvent used to degrease parts)
- Explosives 2,4,6-trinitrotoluene; hexahydro-1,3,5-trinitro-1,3,5-triazine; octahydro-1,3,5,7-tetranitro-1,3,5,7-tetrazocine
- Chlorophenol, PAHs (1–octadecene; 2,6,10,15,19,23-hexamethyl-

tetracosane; phenanthrene; fluoranthene; and pyrene), and Aroclor 1232
- Mineral oil and grease, diesel, JP-4 (jet propellant), and gasoline
- Almost any hydrocarbon-based material

See "Bio-Remediation—The Natural Way" at http://www.naturesway resources.com/resource/infosheets/bioremediation.html for more information.

Making Your Own Mulch

Over thirty states in the United States have now passed laws restricting the amount of organic materials entering landfills, hence disposal is becoming an increasing problem in some areas.

If one has a shredder or chipper, it is easy to chop up woody branches up to 2–3 inches in diameter. These chips can be used as mulch or a carbon source for making compost. These chips (native mulch) make one of the most beneficial mulches. If you have a large amount of brush or large pieces from tree removal, these can be stockpiled, then periodically you can have a grinding company come in and grind the material for you, as large commercial grinders are very expensive to own and operate.

One of the easiest ways to get rid of leaves is to use a lawn mower with a mulching blade (works best if the blade is sharp). One simply mows over the leaves, and they are chopped up into small pieces that will settle between grass blades and decompose quickly, improving the organic matter content of the soil. Mowing is often cheaper and faster than raking and disposal when all costs are included. Dry leaves can be very abrasive, so on large areas check the blade periodically to ensure that it is still sharp.

Several companies now offer leaf-bagging equipment that will vacuum up the leaves, which can then be used in a compost pile. Many landscapers will also pay a dump fee to get rid of their leaves and brush trimmings, which could be another source of material. Many tree service companies will also provide chipped branches and limbs. Small amounts of material are usually not a problem, but if one accepts larger amounts or collects a dump fee, then your state regulations may require some form of permit. Always check with your state environmental or solid waste departments first.

How to Calculate the Amount of Mulch Needed

See Table 7.1 for an explanation of how to correctly calculate the amount of mulch needed for a particular space.

Table 7.1. How to Calculate the Amount of Mulch to Use
Application Rates for Loose Mulch (Not Compacted)

Thickness	*Square Feet*	*Mulch Required*
1"	100	0.3 cu. yd.
2"	100	0.61 cu. yd.
3"	100	0.91 cu. yd.
4"	100	1.11 cu. yd.
6"	100	1.85 cu. yd.

1) To calculate amounts needed for areas with different square footage, multiply the number of square feet by 0.0031 (exactly 0.0030864) cubic yard per square foot for a layer 1 inch thick. Then multiply that number by the number of inches of mulch desired.
2) Another way to estimate mulch or compost needs can be expressed as:

   Volume (cubic yards) = square feet × depth (inches)=324

3) A third way is to remember that 1 cubic yard of product will cover 324 square feet 1 inch deep.

The Science of Mulch

Often we hear people talk about the possibility of "nitrogen tie-up," or robbing the soil of nitrogen, when using mulch, thereby depriving the plants of a required nutrient. In practice, this can only happen when proper mulching techniques are not followed or a low-quality mulch is used. It is important to remember that even when nitrogen tie-up does occur, it is only temporary, for in time the nitrogen is returned to the soil. In fact, extra nitrogen is usually returned to the soil.

Healthy soil with low insect and disease pressure has a ratio of carbon atoms to nitrogen atoms that is 30 to 1. If we incorporate (till or mix) a carbon-rich material into the soil, it can temporarily cause a nitrogen tie-up problem (i.e., sawdust or pine shavings can have a 300:1 carbon-to-nitrogen ratio or higher) until microbes break down the mulch into humus. The key words here are *to till* or *to mix*, which means we are not using the material as mulch but as a soil amendment. Soil amendments have a different set of rules that need to be followed for plants to be healthy. In practice, if a mulch material is placed on top of the ground, it rarely causes nitrogen tie-up problems except for special cases (see the mulch material descriptions below for more specific risk factors).

As we learn more about the importance of soil life-forms and the role they play in producing healthy, disease-free plants, we are also learning that these microbes are sensitive to certain chemicals. For example, the bisulfate anion (HSO_3-), which is a solubility product of sulfur dioxide (SO_2), is highly toxic to soil bacteria and fungi, and these chemicals may be found in acid rain.

As the soil becomes acid, metals like aluminum and to a lesser extent manganese are mobilized and can become toxic to microbes or have negative effects (e.g., prevent fungal spores from germinating). Hence, one needs to be very careful with (or avoid) chemical fertilizers or additives such as aluminum sulfate that can lead to microbe toxicity. It is just safer and easier to use natural or organic ingredients to amend soils.

Also, as a soil becomes more acid, it tends to favor and lead to an increase of fungal microbes over bacterial, which may affect the growth of some plants. Remember that some plants prefer bacteria-dominated soils.

A five-year study at Pennsylvania State University has revealed that colored polyethylene sheeting can increase yields as much as 25 percent when compared with black plastic. Each crop was found to have its own preference. Plant response is believed to be related to the way each color of plastic reflects light and heat (*Horticulture*, January 1997). Additional research has been done at Clemson University, Pennsylvania State University, the USDA, and other institutions. Some of what is being learned is outlined below:

Peppers—performed best when mulched with yellow sheeting
Tomatoes—red plastic worked best
Strawberries—with red plastic, they ripen more quickly, emit a stronger aroma (90% increase in aromatic compounds over black plastic), are almost 20 percent larger, and have higher sugar and organic acid concentrations as compared with black plastic mulch
Squash—blue and red plastic worked best
Silver—aphids tend to ignore plants mulched with silver plastic
Orange—turnips grow bigger
Blue—turnips had a sharper taste
Green—turnips had the sweetest taste (chemical analysis confirmed the most sugars)
White—plants have a thicker wax coat thus use less water
Yellow—attracts the greatest number of insects

Other research has found that the frequency or color of light reaching plants affects the growth of disease organisms as well as plant growth. The spores from the pathogen *Botrytis* require ultraviolet light to germinate. Several fungal diseases require blue, ultraviolet, or infrared radiation to multiply. Research at Disney World's Epcot Center has found that red light reduces foliar fungal diseases on tomatoes, peppers, and cucumbers. It also found that blue wavelengths stimulated the development of these diseases. However, in some cases, the elimination of certain frequencies also reduced plant growth.

Many plant responses to light reflected from colored plastic mulch are dependent on the exact species and cultivar tested. This means that a different species of pepper may do worse with a given color than another. Also, it has been found that two red plastic mulches that look identical to the human eye have very different plant growth responses, as they reflect light differently at wavelengths invisible to humans. It turns out that plants are sensitive to radiation in the far-infrared and ultraviolet wavelengths that humans cannot see. It should be noted that the performance attributes listed in the research reports only compare colored plastic mulch to black plastic mulch.

This research into the reflective properties of plastic mulch explains why straw or hay and fresh ground wood chip mulches work so well. These mulches are very light in color and reflect all colors or wavelengths of light. This allows the plant to use all the colors or whichever color is best. They also enrich the soil as they break down.

As scientists continue to study the effects of light on plants and diseases, we will learn how to use the reflective nature of different types of mulch to its fullest. Plastic mulches have their greatest benefits when used in commercial agriculture to help warm the soil in the spring. A lot of research on plastic mulches is in progress, and we will learn how to use them more effectively in the future.

Effects of Colors

This is a new area of research that we are beginning to understand. The color of mulch or other materials affects plant growth. Research at Texas A&M University studied trees that were planted in paving bricks (pavers) of three different colors: a light (blond), medium (red-brown), and dark (charcoal). The light- and medium-colored bricks reflected the most photosynthetically active (growth-promoting) radiation. The air temperature above the plants was less for the lighter colors as compared with the darker colors. On sunny days in the fall and winter, the air temperature was as much as 35°F higher, which could make the trees (plants) more susceptible to damage from sudden cold snaps. In addition, the darker the color, the more the root growth decreased in the upper portions of the soil, which resulted in reduced growth aboveground. This effect was more pronounced in the shallow-rooted species.

Genetic Effects

The USDA has found that a plant's gene activity changes with the type of mulch applied and with the type of fertilizer used. Some of the research was done on tomato plants, and they found that at least ten different genes were affected. For example, when an organic mulch like mown hairy vetch was used instead of black plastic, the tomato plants lived longer and developed less fungal disease. When the organic vetch mulch was used, two genes for plant defense (immune system) and two genes for regulation of aging greatly increased their activity. The researchers also found that vetch-mulched fields receiving only half as much fertilizer produced larger yields than conventional plastic-mulched fields with the full amount of fertilizer. The vetch-mulched fields also provided other benefits such as reduced erosion, decreased disease, and delays in plant aging. One of the genes studied produces chitinase, an enzyme that dissolves the walls of attacking fungi, along with osmotin, another defensive compound, and extra activity of receptors for cytokinins that regulate plant aging. The mowed-vetch-mulched tomatoes developed root systems that allowed for better nutrient absorption.

Composted Mulches

Composting mulch at high temperatures for a period of time is one of the best ways to increase the quality of the mulch and eliminate potential problems before they occur. However, there are best management practices that the processor needs to follow to ensure high quality.

When large piles of fresh material are composted correctly, the temperatures in the piles rise to 140°F–160°F or more. These high temperatures favor microorganisms known as thermophiles (primarily facultative aerobes) that love the heat (and actually produce it) and perform the initial decomposition called composting. These microorganisms die when the mulch is placed in the landscape and cooled off to 50°F–80°F. Since they require high temperatures to survive, they cannot grow or compete with soil microorganisms. If mulch is taken straight from a high-temperature processing pile and applied to the landscape, the lower temperatures often will create conditions called a "biological vacuum." A biological vacuum can also occur from bagging mulch or from letting it get too dry before application. These can cause problems after the mulch is applied or bagged, allowing the wrong type of microorganisms (pathogens) to grow and colonize in the mulch.

If the piles get too hot, then the range of beneficial microbes at work become very restrictive and phytotoxic acids are often produced. Some producers use techniques that produce very high temperatures because it speeds up the composting process, but it also results in lower product quality.

Potential problems can be prevented in several ways. First, the producer can move the mulch to smaller piles and allow it to cool off for a few weeks before selling the product (a process known as "curing"). This allows the beneficial soil-dwelling microbes known as mesophiles to colonize the mulch before application. Many producers screen their mulch before selling it. The screening process cools off the mulch, which is then placed in smaller piles for sale. The second method is to soak the mulch with water, after application, to help it settle in and to speed up the colonization by beneficial microbes. The addition of water encourages beneficial bacteria to grow and colonize the material, bringing the fungal and bacterial species into a healthy balance (fungus can grow in drier conditions than bacteria). Often mulch producers allow mulch to dry out, which reduces the weight, allowing more product to be delivered on the same truck and thus reducing the delivered cost per cubic yard.

Note: We also receive an environmental benefit when compost is used. In the United States today, we have dead zones along our coastlines caused by excess nutrients from agricultural fertilizer runoff. Some of these dead zones are from New Orleans to Galveston along the Gulf Coast and from Baltimore into Virginia and Chesapeake Bay. When we compost manure and other organic materials to make mulch or compost, we significantly improve the quality of water entering the nation's watersheds, as they do not leach nutrients as much as artificial fertilizers do.

The Mulch Business

Mulch Naming Conventions

Often we find many vendors of mulch products using the same terminology to describe vastly different products. The consumer often ends up confused or disappointed when the product arrives. In an effort to standardize product descriptions and prevent fraudulent claims, the Mulch and Soil Council (formerly the National Bark and Soil Producers Association) has devised the "Voluntary Uniform Product Guidelines for Horticultural Mulches, Growing Media and Landscape Soils," which can be downloaded from their Web site at http://www.mulchandsoilcouncil.org/certification/standards.php. The following are some of their definitions for mulch products:

Pine Bark Nuggets: Products derived from the genus *Pinus* with particle size from 1.25" to 2.75" in diameter and a maximum wood content of 15 percent.

Pine Bark Mini-Nuggets: Products derived from the genus *Pinus* with particle size from 0.50" to 1.25" in diameter and a maximum wood content of 15 percent.

Pine Bark Mulch: Products derived from the genus *Pinus* with a maximum wood content of 15 percent.

Western Bark Mulch: Products derived from conifer trees common to the western region of North America and with no more than 15 percent wood content.

Western Bark Nuggets: Products derived from conifer trees common to the western region of North America with particle size from 1.50" to 2.25" in diameter and a maximum wood content of 15 percent.

Western Bark Mini-Nuggets: Products derived from conifer trees common to the western region of North America with particle size from 0.75" to 1.50" in diameter and a maximum wood content of 15 percent.

Hardwood Bark Mulch: Products derived from deciduous hardwood trees and with a maximum wood content of 15 percent.

Cypress Mulch: Products derived 100 percent from trees of the genus *Taxodium.*

Hemlock Bark Mulch: Products derived from trees of the genus *Tsuga* and with a maximum wood content of 15 percent.

Mulch Products: Consist of 100 percent bark and wood products (excluding the use of any reprocessed wood products), mechanically screened and/or shredded, containing at least 70 percent of the named material as follows:

Pine Mulch: Products derived 100 percent from conifers of the genus *Pinus.*

Hardwood Mulch: Products derived 100 percent from deciduous hardwood trees.

Cypress Mulch: Products derived 100 percent from trees of the genus *Taxodium.*

Western Mulch: Products derived 100 percent from conifers common to the western region of North America.

Cedar Mulch: Products derived 100 percent from trees of the genus *Thuja* or juniperus.

Stump and Root Mulch: Products derived 100 percent from the processing of tree stumps and/or roots.

Mulch Blends: Consist of bark, wood products, or reprocessed wood products containing more than one genus or a mix of forest products and/or reprocessed wood that have been mechanically screened and/or shredded. If reprocessed wood products are used in any portion of a blend, such use must be indicated on the product ingredient label and shall not contain hazardous materials above limits established by the Environmental Protection Agency (EPA).

Wood Mulch: Consists of wood, wood products, or reprocessed wood that is mechanically shredded and/or screened and shall not contain hazardous materials above limits established by the Environmental Protection Agency (EPA).

Mulches Not Recommended

Rubber Mulch

Rubber mulch is typically made from ground-up recycled tires and has generated a lot of discussion on the benefits and risks or dangers of using it.

There are two schools of thought on rubber mulch. It seems that all the studies paid for by the rubber mulch manufacturers and tire companies show benefits after their PR firms get through with them, while all the independent studies show that it is toxic and dangerous.

Let's look at each of these claims and see what the research says.

Doesn't float away, doesn't blow away, doesn't sink into the ground

Most rubber mulches have a specific gravity greater than water, hence they do not float or wash off as easily as some other materials like bark mulches. If the soil is healthy and full of microbes and earthworms, *all* materials will settle and sink eventually. If a material is heavier (denser) than water, it will sink or settle faster than a lighter material like an organic mulch.

Doesn't decay away; lasts many years

Rubber mulch is broken down by microbes like any other product (remember, microbes can break down granite rocks into soil); rubber is easy by comparison. The rubber mulch encourages species of bacteria that break down rubber and rubberlike products in your home to multiply. The additives in tires to prevent bacterial decay (which are toxic chemicals) are broken down by white and brown rot fungal species that live in soil. This same decomposition is what releases the toxic chemicals in tires.

Doesn't feed or house insects

Nothing eats tires except microbes. However, the tire mulch does kill many species of good microbes that kill insects and prevent disease. The toxic chemicals in the tires will also kill beneficial insects that help control pests.

Doesn't smell, mat, or mold

This may be a matter of opinion, but most people find that rubber mulch starts to stink as it gets hotter. On a hot day, it has a strong stench. As tires are ground up into chips, the amount of surface area is greatly increased and all the new surfaces are freshly exposed, allowing for maximum odors to be released. Some people get sick from just being in the sales area of a store selling new tires.

As rubber mulch heats up, it releases toxic gases such as volatile organic compounds (VOCs) and another class of chemicals called polycyclic aromatic hydrocarbons (PAHs). These gases have been found to cause irritation of the nasal and respiratory passages, central nervous system damage, depression, headaches, nausea, dizziness, eye and kidney damage, and dermatitis. Hence, ground tires should never be used in an enclosed area or indoors. These effects would be even worse in areas of high air pollution (e.g., Houston).

It is true that rubber mulch will not mat down as easily as organic mulches, since the beneficial microbes that create soil structure and prevent insect and disease problems cannot live in it.

As to mold, I have seen many tires used on piers and boat docks covered with algae. I have also seen tires used as planters covered with what appears to be mildew and mold. Mildew and mold will grow on about any surface if moisture is present, unless it is too toxic and something kills them.

Impedes weed growth

In comparison studies of several mulch types in herbaceous perennial beds, rubber tire mulch was less effective than even raw wood chips. Other studies have found that even sawdust worked better and that rubber mulch was less effective than straw and other fibers in impeding weed growth.

Several studies have found that rubber tire mulches kill many species of plants. The public relations specialist can spin this effect as "retards or impedes weed growth." Who wanted flowers in the first place?

Also, as temperatures rise, the types of plants that will survive are reduced—and rubber mulch can get fairly hot (see below). Metal toxicity also reduces the types of plants that can live and grow in rubber mulch (see below).

Allows water and nutrients to permeate

Some researchers have found that ground-up tires can absorb chemicals from fertilizers and pesticides, preventing them from leaching into groundwater. As a result, however, fewer nutrients reach the plants. Eventually, the tire chips will degrade and the stored chemicals will be released, most likely at time when the plant does not want or need them.

Water will indeed run through the tire mulch, as it is highly permeable. However, the problem begins when the water reaches the surface of the soil. Earthworms and soil microbes create a soil structure that allows air and water to enter the soil. But earthworms and microbes require decomposing *organic* matter as a food and energy source. Without this food source, most of the earthworms and microbes will die off and the soil structure will collapse over time. When this happens, the soils will become anaerobic (which favors root pathogens), and water and air cannot enter easily. Hence, conditions may be created that favor disease, and since the water can no longer be absorbed, it must run off.

Safe for flowers, plants, and pets

Research at Bucknell University has found that the leachate from ground tires can kill entire aquatic communities of algae, zooplankton, snails, and fish. Even at low concentrations, it can cause reproductive problems and precancerous lesions. Also, marine life, from seaweeds to plankton, is negatively affected.

The toxic nature of the leachate from tire rubber is due at least in part to the chemicals used in producing tires (cadmium, chromium, aluminum, copper, iron, manganese, molybdenum, selenium, sulfur, and zinc). Of these minerals, rubber tires may contain extremely high levels of zinc, even up to 2 percent of the tire mass. Many plant species have been shown to accumulate zinc in their tissues to the point of death. USDA researchers who have studied the effects of metals in sewage sludge, biosolids, and compost have found that ground rubber should not be used for any agricultural or garden soil, potting media, or compost. Yes, some companies use tire chips and crumb rubber as a bulking agent for compost—and we wonder why the compost does not work and is toxic to plants!?

Other rubber leachates have been found to cause problems from skin and eye irritation to major organ damage and even death. Long-term exposure can lead to carcinogenesis and mutagenesis. For example, 2–mercaptobenzothiazole used in vulcanizing rubber is highly persistent in the environment and harmful to aquatic life. Ground rubber also contains a class of chemicals called polycyclic aromatic hydrocarbons (PAHs) that many studies have found extremely toxic to humans and the environment. Research has also found that the toxicity of leachate from the rubber tire mulch increases over time as the rubber breaks down.

Colored Mulch

Colored or dyed mulch is a product that comes in many forms and has evolved and changed a lot over the last few years. Many of the early problems have been overcome, but it is still a special-purpose mulch with limited uses, and *must be used correctly* or it may result in stunted plant growth coupled with increased insect and disease problems.

Production of Colored Mulch

Colored mulches are made by grinding up some form of dry wood waste into chips and then dying them with a water-based solution. For the dyes (colorants) to stick to the wood, it has to be dry and some form of chemical binder needs to be used.

History of Colored Mulch

Colored or dyed mulch entered the market several years ago and created many problems for consumers. The inexpensive early dyes often contained toxic heavy metals and other contaminants in the colorants. The natural or organic dyes were far more expensive and hard to find, did not work as well, faded quickly, and became unsightly. Many of these problems have now been solved, but risks remain.

Drawbacks of Colored Mulch

Colored mulch is not widely available, but availability is improving. The colored mulches made with organic dyes are more expensive than the ones made with the toxic chemical older types of dyes. Colored mulches do not support the types and variety of beneficial microbes required for healthy plants as do other organic mulches, such as a good native mulch or compost.

To absorb the dye, the mulch has to be made from fresh dry wood. This may be unused construction wood scraps, mill waste, or old pallets. Some vendors grind up and use the hazardous CCA-treated wood, so the colored mulch may be an unwanted source of arsenic (studies in Florida have found levels of arsenic in colored mulches to exceed federal safety limits in over 75 percent of the samples tested).

The preservatives used in the colorant may also inhibit the growth of many beneficial soil microbes that prevent disease in plants. The dry wood waste and old pallets also have a very high carbon-to-nitrogen ratio (200 to over 500:1), hence if supplemental nitrogen is not added, it may cause a nitrogen tie-up in the soil. Healthy fertile soil has a C:N ratio of only 30:1, so nitrogen is pulled from the soil as microbes try to break down and decompose the colored mulch. This often results in poor plant growth with increased disease and insect problems.

If old pallets were ground up, the colored mulch may contain dangerous chemicals, depending on what the pallets were used for. The colored dyes are often used to mask or hide the grayish color associated with recycled C&D (construction and demolition) waste wood.

Researchers at Penn State have found that artillery fungus is attracted to and often grows on colored mulch made from old pallets. Artillery fungus causes black spots on the sides of houses, cars, and other structures that are very difficult to remove. Some colorants will delay the onset of artillery fungus.

Mulch from the type of wood required to make colored mulch often becomes hydrophobic when dry and will not absorb water easily, hence plants dry out and become stressed or even die.

The vendors of some colored mulches use powdered dyes, which quickly rub or wear off, and the customer is left with an ugly mess. These are more commonly found in the cheaper colored mulches. Dust from the powdered dyes is easily inhaled, causing respiratory irritation.

Colorants are expensive, and if the vendor does not use enough colorant, then the color will quickly fade and look very unsightly. In the best cases, the colorants only last one year and less in warm, humid climates.

Black colorant such as carbon black may be made by the incomplete combustion of petroleum products (possibly carcinogenic to humans, and short-term exposure to high concentrations of the carbon-black dust may produce discomfort to the upper respiratory tract through mechanical irritation). Higher temperatures increase the release of chemicals into the environment (black colors absorb more energy from sunlight, hence have a higher surface temperature). Also, nitrates (i.e., synthetic fertilizers) increase the breakdown and release of toxic chemicals.

Natural carbon black from vegetable origin is used as a food coloring (in Europe it is known as additive E153). Carbon black (PBL-7) is the name of a common black pigment traditionally produced from charring organic materials such as wood or bone. Ask the mulch vendor for a Material Safety Data Sheet (MSDS) to be sure which one is used.

Blue, green, and yellow are other colors that are available. Very little information regarding how or what they are made from has been published, and no MSDS reports were found except for one yellow colorant made from a hydrogenated iron oxide compound that is relatively safe. A lack of published information is a clue that the manufacturers of the chemicals do not want the customer to know what is in them.

Advantages of Colored Mulch

Many of the new dyes are natural, such as iron oxide for a red color or carbon black to get a black color. Iron is a plant nutrient, and the red mulch will provide small amounts of iron as it breaks down.

The new water-soluble spreader stickers—additives used to make the dye cover the particles evenly and adhere better—are biodegradable, thus many of the earlier problems of chemical binders can be avoided. These are similar to the ones used to help compost tea, seaweed, and fish emulsions stick to plants when applied.

The colored mulches can be very attractive when used in the correct setting and the proper manner.

- Green mulches have been used on concrete floors to protect horses and other animals for rodeos and livestock shows.
- Black mulches have been used to absorb energy from the sun to warm animal pens.
- Colored mulches can be used on pathways where one does not want plants to grow, as they are slower to break down compared with more natural mulches. The negative effects also reduce weed growth.
- They can be used to accent the color of flowers planted adjacent to the pathways (e.g., greenish gray or soft green mulch will make the yellow flowers of daylilies stand out).
- Colored mulches can give the "look" of redwood mulch, and their use helps the environment, since valuable redwood trees are not destroyed *if* the mulches are produced properly.
- They can be used on the floor at trade shows or exhibits and for other display purposes and can be very attractive when used in this manner.
- They can be used to cover a bare area and then set potted plants or patio furniture on in a townhome or other small space where a flower bed is not wanted.

- They are lightweight as compared with other mulches, hence easy to move, handle, and apply.
- They are often available in bagged form for convenience.
- They are offered in various colors and used for their aesthetic appeal.
- They may be used for their reflective properties, similar to colored plastic for some plants.
- They are made from recycled materials and a good use of material that would have been burned or dumped, using up valuable landfill space.

Correct Usage of Colored Mulch

If one chooses to use colored mulch for flower beds, only a thin layer should be used (½–1 inch thick) over a 2–3-inch-thick layer of good compost or native mulch. The layer of compost or native mulch acts as a buffer, minimizing the negative effects of the raw wood. The good mulch layer gives the soil and plants the health benefits they require, and the colored mulch gives the aesthetic look one may want.

A possible use for colored mulch is for its reflective properties, but no controlled studies have been found (a future research topic). What we have learned is that red plastic has been found to increase yields (12–20%) and sweetness of tomatoes when compared with other colors. Strawberries grown with red plastic ripen more quickly, emit a stronger aroma (90% increase in aromatic compounds over black plastic), were almost 20 percent larger, and have higher sugar and organic acid concentrations as compared with those grown with black plastic mulch. However, shallow root systems are often created by all plastic mulches, and, during drought periods, the plants may not survive the stress. Other drawbacks are the same as for clear or black plastic. Colored wood mulches might give the same benefits without the negative effects. As research into mulches continues, we will get better answers in the future.

As with all products, both the quality and the price will vary for colored mulches. The old saying holds true: "You get what you pay for." If one wishes to use colored mulches, it is best to buy only from a reputable vendor and use it correctly, but we strongly recommend avoiding this product.

CHAPTER 8

Landscaping

Landscaping projects use the Basic Organic Program of soil preparation and fertilization. In most cases, three applications of natural organic fertilizer a year are recommended in the early years. That can be reduced to two after a few years and then one application a year after healthy soil with reasonable biological activity has been established. Replenishing the carbon that is removed from the soil is the key element in landscape fertilization.

Color beds usually require the most amendments because they are primarily annual plantings that are removed after flowering (and hopefully deposited in the compost pile to later be recycled into the same project). In most cases, these plants come from nurseries where the plants are grown with daily inputs of water and chemicals and must be weaned off of intensive feeding and adapt to real-world conditions in the field. Another reason more inputs are needed is that the annual plants are removed after their seasons end, and along with the plants go nutrients that have been pulled up by the plants. Applying foliar feeding with Garrett Juice and other organic sprays helps these plants adapt to the more rigorous conditions outside the nursery. Building the soil with not only the Basic Organic Program fertilizer recommendations but also an addition of either fish meal, alfalfa meal, or some other organic amendment is very helpful. We have also found that the addition of natural diatomaceous earth and/or zeolite to the annual beds is highly beneficial.

Landscaping Elements

Bed Preparation

Into the native soil, till 3–6 inches of quality compost and the following amendments per 1,000 square feet: lava sand at 80–150 pounds, greensand at 40 pounds, dry molasses at 10–20 pounds, cornmeal at 20 pounds, and

Trunk flares exposed—one of the most important goals of proper tree-planting techniques. Photograph by Howard Garrett.

organic fertilizer per label directions. Till the amendments into the existing soil, creating raised beds. In low-calcium soils, replacc the greensand with 80 pounds of high-calcium lime per 1,000 square feet. Removal of existing soil prior to adding amendments is not recommended. Unless it would block surface drainage, amendments should be added above the native soil to create a raised bed. Removing soil and filling beds with new material causes bathtub effects and severe drainage problems in many cases.

Fertilizing

Apply a natural organic fertilizer two to three times per year. During the growing season, spray turf, tree and shrub foliage, trunks, limbs, and soil monthly with compost tea, molasses, natural apple cider vinegar, and seaweed mix (Garrett Juice; see Appendix 1). Add volcanic and other rock mineral sands once every two to ten years at 40–80 pounds per 1,000 square feet. These rock minerals are no longer needed after soil health has been obtained.

Mulching

Mulch all shrubs, trees, and groundcovers with 1–3 inches of compost, shredded tree trimmings, or shredded hardwood bark to protect the soil, inhibit weed germination, decrease watering needs, and mediate soil temperature. Use 1–2 inches of compost under mulch to duplicate the forest floor profile. Shredded native tree trimmings make the best mulch of all in most cases. Small bedding plants and herbs can be mulched with compost or other fine-textured mulch.

Mulching is the greatest water-conserving, soil-conserving, and soil-building practice possible. It breaks up heavy water droplets, stopping them from splashing the soil loose. Mulch holds the water in place and keeps the sun from heating up the soil or the cold winters from freezing it. Mulch keeps the soil temperature and moisture more even and slows evaporation. This encourages all types of beneficial microbes, fungi, and actinomycetes to churn up the soil so the water will penetrate instead of running off to cause soil erosion and flooding. Shredded tree trimmings become fibrous when ground, stick in place, don't wash away, decay very slowly, and don't tie up plant nutrients, especially nitrogen, while they decay.

Mulching the bare soil in landscaping provides numerous benefits:

- Conserves moisture
- Maintains even soil temperature
- Increases root zone
- Stops erosion
- Prevents crusting—increasing water absorption and aeration
- Helps control weeds
- Keeps soil from splashing on lower leaves and vegetables
- Provides walkways—your feet don't get muddy
- Recycles waste materials (tree trimmings, leaves, etc.)
- Acts as buffer against high/low pH and overfertilizing and increases soil exchange capacity
- Increases the numbers and activity of all beneficial soil life (microorganisms and earthworms), and their activity is of utmost importance to healthy soil and plants

Some of the known benefits of soil life are:

- Growth-promoting hormones are made
- Toxins and antibiotics are made that protect plants from soil diseases
- Sticky substances are created that glue soil into crumb texture, creating aeration
- Many types of natural acids are formed to help unlock minerals
- Can detoxify poisoned soils
- Improves plant establishment and increases plant growth (mycorrhizal fungi)
- Improves plant nutrient and water uptake (mycorrhizal fungi)
- Breaks down organic materials into nutrients and slowly releases them back to the plant
- Some beneficial soil microbes destroy harmful insects (nematode-destroying fungi)
- Some microbes have the ability to fix nitrogen from the air while breaking down high carbon materials

Watering

Adjust schedule seasonally to allow for deep, infrequent watering in order to maintain an even moisture level. Start by applying about 1 inch of water per week in the summer and adjust from there. Water needs will vary from site to site.

Mowing

Mow weekly, leaving the clippings on the lawn to return nutrients and organic matter to the soil. General mowing height should be 2½ inches or taller. Put occasional excess clippings in compost pile. Do not bag clippings unless it has to be done to move the material to the compost pile. Do not let clippings leave the site. Do not use line trimmers around trees. Mulching mowers are best if the budget allows.

Weeding

Hand-pull large weeds and work on soil health for overall control. Mulch all bare soil in beds. Avoid synthetic herbicides, especially preemergent broadleaf treatments and soil sterilants. These are unnecessary toxic pollutants. Spray broadleaf weeds with full-strength vinegar. Add 2 ounces of orange oil and 1 teaspoon of liquid soap per gallon for additional power. Fatty-acid products such as Scythe and Monterey Insecticidal Soap can also be used.

Pruning

Remove dead, diseased, and conflicting limbs. Do not overprune. Do not make flush cuts. Leave the branch collars intact. Do not paint cuts except on oaks in oak-wilt areas when spring pruning can't be avoided. In those cases, use natural shellac or Lac Balsam.

Landscaping Specifications

Specifications are the written instructions on how to prepare beds, install the plants, and maintain the plants. They are needed on residential and commercial projects. They can be simple or very detailed and complicated. The following is an example of specs that would be used for organic maintenance. Installation specifications would be done in the same form with planting details added. Details are covered in Chapter 3.

Commercial organic project—Texas Discovery Gardens at Fair Park, the State Fair grounds in Dallas, "the first public garden in the state of Texas to be certified 100% organic by the Texas Organic Research Center" (http://texasdiscoverygardens.org/the_gardens.php). Photograph by Howard Garrett.

Organic Landscape Maintenance Specifications

Part 1: General

1.1. Scope

A. Furnish all work and materials, appliances, tools, equipment, facilities, transportation, and services necessary for and incidental to performing all monitoring, adjustment, and minor repair of sprinkler irrigation system; irrigation scheduling; inspection of drainage system; weeding of mulched beds; mulching of beds; pruning of trees; pruning of shrubs; applications of fertilizers, insecticides, and herbicides; general site cleanup; removal of trash and products of maintenance; and submittal to owner of maintenance schedule. When the term *contractor* is used in this section, it shall refer to the Landscape Contractor.

B. Related Work Specified Elsewhere
1. Landscaping
2. Landscape Irrigation System

1.2. Requirements of Regulatory Agencies

Perform work in accordance with all applicable laws, codes, and regulations required by authorities having jurisdiction over such work and provide for all permits required by local authorities.

1.3. Contractor Responsibilities

A. The contractor shall be responsible for all planting beds, all plants, and the irrigation systems.

B. The contractor's maintenance period shall be__________________.

Part 2: Materials

2.1. Materials

A. No synthetic fertilizers or toxic chemical pesticides are allowed.

B. Only products acceptable to the Texas Organic Research Center (TORC) are allowed to be used.

2.2. Fertilizers (all plantings)

A. Fertilizer samples to be submitted for approval by the owner prior to use.

B. Acceptable fertilizers include such products as Lady Bug Natural Brand, Medina, Maestro Gro, and GreenSense.

2.3. Herbicides

A. Contact postemergent herbicides: vinegar, fatty-acid, and other natural products.

B. Preemergent weed control: corn gluten meal.

2.4. Wound Paint

A. Wound dressings and pruning paints should not be used on pruning cuts.

B. Injuries to trunks and limbs should be treated with Tree Trunk Goop.

2.5. Machinery and Equipment

Machinery requirements listed under this section are NOT intended to be restricted to specific manufacturers or models unless so stated. Specific mention of the manufacturers is intended as a guide to illustrate the final product of the maintenance operations desired. All equipment used shall be maintained in top working condition at all times.

A. Pruning tools shall be maintained in safe working condition; cutting edges shall be sharp at all times.

B. Granular material spreaders shall be the cyclone type. The contractor shall be responsible for any grade, plant material (trees, shrubs, etc.), or hardscape amenity damage caused by the spreader and the application process. Spreaders shall be in a safe working condition at all times.

C. Pest control sprayers for application of organic products shall be of the handheld or backpack type. The contractor shall be responsible for any grade, plant material (trees, shrubs, etc.), or hardscape amenity damage caused by the sprayer and the application process. Sprayers shall be in a safe working condition at all times.

D. All carts, wheelbarrows, and similar wheeled conveyances used in or on any portion of the existing landscape or amenities shall be equipped with pneumatic tires.

Part 3: Execution

3.1. Watering

A. General

1. Maintenance procedures should assure the operation of the irrigation system. The irrigation system components (valves, nozzles, and controller) should be inspected, cleaned, repaired, and adjusted weekly.
2. The system's timing shall be adjusted in accordance with the general weather conditions. Any plant material whose good health and appearance declines due to improper watering procedures shall be replaced with the same plant of equal size and form at the cost of the contractor.
3. Promptly repair any damage to the irrigation system caused by the maintenance operations; vandalism; excavation by others resulting in broken heads, risers, pipe, or other similar damage. Replace with the same part and manufacturer.

B. All planting areas should be watered as necessary to provide the proper moisture levels. Adjust watering practices to match water requirements of species in planting beds. Maintain uniform moisture in all planting areas during the winter months, particularly when a freeze is predicted.

D. Avoid over- and underwatering and notify the owner immediately if drainage problems appear.

3.2. Fertilization (all planting areas)

A. Fertilize all shrubs, groundcovers, and flower beds in February, June, and September with organic fertilizer at 20 pounds per 1,000 square feet.

B. February and September applications shall be corn gluten meal. June application shall be Medina, Maestro Gro, Nature's Guide, Lady Bug, or other approved product. Dry molasses can be substituted subject to approval in most cases.

C. All plants shall be sprayed with Garrett Juice or approved aerated compost tea monthly. Problem plants should be drenched with Alpha Bio Thrive.

3.3. Trees

A. Staking and guys shall be removed from trees.

B. Any fire ant mounds around or on top of a tree root zone shall be treated immediately with products as specified. Do not allow the mound to build on the tree trunk, as this will cover the tree root flare and possibly cause injury or death to the tree.

C. Prune only as needed rather than on a regular schedule. Experienced pruning personnel shall carry out pruning.

1. Prune to remove damaged limbs; remove crossing branches; and maintain the natural shape of each species.
2. Sterilize pruning tools with hydrogen peroxide between individual plants.
3. Use no line trimmers or edgers within 15 feet of any tree. Should the need for trimming be necessary near tree trunks, it shall be done by hand trimming only.

3.4. Shrubs and Groundcovers

A. Thin to remove dead wood when necessary. Remove dead wood and freeze-damaged leaves in the spring.

B. In no case should any shrub be sheared. Shrubs shall be selectively pruned.

C. All water sprout and sucker-type growth shall be pruned and trimmed continuously. Pruning and trimming of any shrub shall be done in a manner so as to retain the natural character and habit of the plant. All shrubs shall be pruned to create a uniformly dense plant. Selectively thin and tip back annually, or as needed. Do not change the natural shape of the shrub by pruning unless so directed by the owner.

D. All shrub and groundcover beds shall be edged, weeded, and cultivated as needed.

E. Prune out dead, broken, and diseased wood.

F. All damaged, dead, or thin areas in groundcover beds shall be replanted. Replacement of plant material not due to the contractor's negligence will be at the owner's expense, upon receipt of written authorization to proceed.

G. All pruning debris and limbs shall be removed completely and immediately from site or to an approved location on-site.

H. All groundcover beds shall be sheared one time per year. This shall be done in the early spring, prior to the growing season. Groundcover beds bordering on paved surfaces must be edged as needed to retain a neat edge.

I. Remulch beds as necessary to maintain a 2–3-inch depth. Do not pile mulch on the stems of plants.

3.5. Turf Areas

A. **Mowing:** Each mowing, where possible, shall be performed at an oblique or ninety-degree angle to the previous mowing. Corrective height adjustments shall be made on mowers as weather conditions dictate. Never scalp the turf areas or cut more than one-third of the existing top growth in one mowing. Mowers shall be kept sharp for an even cut. If a full mowing is not necessary, spot mowing shall be performed as needed to maintain a crisp and neat appearance. During periods of cool weather, mow as low as 1½ inches, but during hot weather, the cut should be no lower than 2 inches from the soil. Mow weekly. Buffalograss should be mowed once a year at 3 inches. Native meadow grasses shall be mowed twice a year as directed. Other grasses should be mowed as directed.

B. **Edging and Trimming:** All turf perimeters and areas around walks, curbs, walls, bed edging, utilities, and other fixtures shall be edged and trimmed at each mowing or at intervals sufficient to maintain a crisp and neat appearance. Absolutely do not use line trimmers around trees and shrubs. Sprinkler heads shall be trimmed as often as necessary to keep operating properly. The hard surface areas adjacent to turf shall be swept and cleaned after each operation.

C. **Watering:** Provide a regular deep-watering program. The established turf should not be kept wet but should dry out somewhat between watering. A twice weekly watering is good under regular conditions, but if it is hot or windy, water more often. In shaded areas created by trees or shrubs, water more frequently because of the competition for soil moisture. If the turf areas wilt (look gray-brown), water more frequently. If there is no rainfall, water at the rates of ½ to 1 inch per week during the spring, 1½ to 2 inches per week during the summer, 1 to 1½ inches per week during the fall, and ½ to 1 inch every two weeks during the winter.

D. **Fertilization:** Fertilize all turf areas with organic fertilizer as explained under Shrubs and Groundcovers (3.4).

E. **Weed Control:** Control of weeds and undesirable grasses shall be with the use of both preemergent and postemergent products. Before such applications are made, the turf should be well established and in a vigorous condition.

F. **Insects:** Control insects with regular applications of approved products as needed. Use the insect controls as directed in the Materials section (2.1).

G. **Diseases:** When diseases first appear, spray with an approved organic disease control as directed in the Materials section (2.1).

H. **Aeration:** All turf areas shall be aerated with a plug-removal-type aerator. Aeration improves fertilizer and water utilization. This is to be done on an additional as-instructed basis. This is not in the base contract. Provide unit price per acre for optional use.

I. **Cleanup:** Cleanup shall include removal of grass clippings from all walks, curbs, decomposed granite pathways, and paving. Remove this to the compost pile or mulch into the turf.

3.6. Maintenance of Meadow Areas

A. **Mowing:** Mow two times per year, in mid-July after the wildflowers have gone to seed and at the end of October. Use a tractor-pulled mower and mow at a height of 6 inches. Trim around all trees before mowing to protect trees from the tractor mower. To prevent ruts, do not mow if the soil and meadow area are wet. Let clippings fall in a meadow area but clean all clippings from hard-surface areas. During the growing season, mow two walk-behind-mower widths next to walks and parking areas weekly for a more manicured appearance.

B. **Edging and Trimming:** All meadow perimeters and areas around walks, curbs, walls, bed edging, utilities, and other fixtures shall be edged and trimmed at each mowing or at intervals sufficient to maintain a crisp and neat appearance. Absolutely do not use line trimmers around trees and shrubs. Sprinkler heads shall be trimmed as often as necessary to keep them operating properly. The hard-surface areas adjacent to the meadows shall be swept and cleaned after each operation. As needed, hand-trim weeds over 3 feet tall back to a height comparable to meadow plants.

C. **Fertilization:** Fertilize all meadow areas with organic fertilizers.

D. **Weed Control:** Control weeds with acceptable products.

E. **Cleanup:** Cleanup shall include removal of clippings from all walks, curbs, decomposed granite pathways, and paving. Mulch clippings into turf or remove to the compost pile.

3.7. Pest and Disease Control

Apply treatments as required for safe control of the particular diseases or insects.

A. Assess level of damage caused by insects and diseases regularly. Minor, visually unimportant damage does not need to be treated, as long as the long-term health of the planting is not affected.

B. Carefully identify any pest that causes significant damage. Do not attempt control until the pest organism has been identified.

C. After identification, choose the least toxic control measure possible. Read and observe all label precautions. If the least toxic control measure is not effective, use the next least hazardous biological or pest-specific control measures. In pest outbreaks, review cultural practices to determine the underlying cause, and correct.

D. Specific directions are as follows:

1. Insects

a. Aphids: Spray orange-oil-based product and release ladybugs. Neem products can also be used. Spray Bio Wash (to clean plants).

b. Armyworms, cankerworms, leaf rollers, tent caterpillars, sod webworms, webworms, and other larvae of moths and butterflies: Release trichogramma wasps when foliage first emerges in the spring. Treat when insects are active between April and September with *Bacillus thuringiensis*. For quick control of heavy infestations, spray any of the citrus-oil-based products.

c. Bagworms: Release trichogramma wasps at spring leaf emergence. Spray Bt products with 1 ounce of liquid molasses added per gallon of spray. Once bags have formed, hand removal is the only solution.

d. Borers: Active borers in trunks can be treated with full-strength orange oil followed by Tree Trunk Goop. To prevent their return, apply the entire Sick Tree Treatment.

e. Cucumber and Other Destructive Beetles: Treat with neem or plant-oil product. Apply beneficial nematodes to the soil. Spray Garrett Juice with 2 ounces of orange oil added per gallon of spray.

f. Fire ants: Treat mounds with orange-oil-based products. Apply beneficial nematodes and horticultural cornmeal for additional control. Dry molasses applied to the entire site at 20 pounds per 1,000 square feet is an effective alternative program.

g. Galls: Normally not a problem. For heavy infestations, spray neem product and apply the Sick Tree Treatment.

h. Grasshoppers: Treat in the spring with Nolo Bait. Treat insects that are feeding with Surround WP or other kaolin-clay or particle-film product. Spray plant-oil products such as EcoEXEMPT.

i. Lacebugs: Treat at first sign of infestation with horticulture oil, plant-oil product, or neem product. Spray Garrett Juice with 2 ounces of orange oil per gallon of spray.

j. Leafminers: Treat with neem or garlic-pepper tea when first symptoms appear on leaves, usually in summer months.

k. Pine tip moth: Release trichogramma wasps in the early spring. Spray Bio Wash (to clean plants).
l. Pine bark beetle: Treat with neem or citrus-based product.
m. Scale: Treat infestations with horticultural oil or orange oil with Bio Wash (to clean plants) added to the spray mix. Follow the temperature restrictions for use of horticultural oil or orange oil/D-limonene insect control product.

2. Beneficial Insect Release: Release ladybugs (2,000 per 1,000 square feet) directly on plants infested with aphids. Adjust as needed depending on the troublesome insect populations. Spray plants with molasses at 2 ounces per gallon prior to release.

a. March–May: Release trichogramma wasps at 10,000 eggs per acre weekly. Begin releases when tree leaves start to emerge.
b. April: Release green lacewings at 2,000 eggs per acre weekly for four weeks.
c. May–Sept.: Release green lacewings at 1,000 eggs per acre every two weeks.

For more information on organic insect control, see C. Malcolm Beck and John Howard Garrett's *Texas Bug Book.*

3. Disease Control: Trees, shrubs, groundcovers, and vines

a. Fungal Diseases
   i. Treat all fungal diseases such as powdery mildew, rust, leaf spot, fungal leaf spot, oak leaf blister, pine needle rust. Spray plants with Garrett Juice plus garlic. Treat soil with horticultural cornmeal at 20 pounds per 1,000 square feet. Spray Bio Wash (to clean plants).
   ii. Oak wilt: Spray Garrett Juice plus garlic tea and apply the entire Sick Tree Treatment.

b. Bacterial Diseases
   i. For all other bacterial diseases such as pine twig blight, pine needle blight, and sycamore anthracnose, spray Garrett Juice plus Consan 20 or 5 percent hydrogen peroxide. Spray Bio Wash (to clean plants).

Note: Other approved compost tea products such as GreenSense Foliar Juice can be substituted for Garrett Juice.

4. Disease Control: Turfgrasses

For fungal diseases such as pythium blight, rust, helminthosporium, take-all patch, fusarium blight, brown patch, and gray leaf spot, spray garlic or garlic-pepper tea and apply horticultural cornmeal or organic whole ground cornmeal at 20 pounds per 1,000 square feet at first sign of disease. Bacterial diseases shall be controlled with

Consan 20 or 5 percent hydrogen peroxide. Spray Bio Wash (to clean plants).

5. Mice, moles, gophers, skunks, rabbits, armadillos, beavers, and other beasts

Each case should be dealt with on an individual basis. Apply the products Rabbit Scram, Mole Scram, Deer Scram, and Gopher Scram as needed. Use snap traps in bait stations for mice and rats.

3.8. Weed Control

A. Preemergent: Apply corn gluten meal at 20 pounds per 1,000 square feet March 1 and October 1.

B. Postemergent: Spot treat annuals and perennial grasses and forbs with vinegar herbicide, Scythe, Burn Out, Black Jack 21, Monterey Herbicidal Soap, or other approved organic herbicide.

CHAPTER 9

Commercial Growing Operations and Recreational Properties

Commercial Growing Operations

Orchards and Tree Farms

Choosing to grow native food crops is not an option for most people. With the exception of figs and jujubes, few fruit and nut varieties can be used throughout the country. That means that we are basically planting relatively ill-adapted plants in most cases. It doesn't mean we can't grow these wonderful food crops, we just have to give the plants more tender loving care because they won't be perfectly happy. That starts with soil improvement. As the budget allows, build the carbon and the rock minerals as discussed in Chapter 2. Plant the trees with the planting techniques explained in Chapter 3 and then use the Organic Pecan and Fruit Tree Program described in Appendix 1.

Tree Farm Marketing

Not only is there a great market for organically grown trees but trees are also the easiest crop of all to grow organically. All sorts of typical problems are eliminated, and the expenses go down in direct proportion. The trees can be sold at a reasonable price, and there will be virtually no losses from transplanting because the root systems will be much more healthy, dense, full of beneficial mycorrhizal fungi, and productive.

Tree Management

Many of the tree problems that are diagnosed by so-called tree experts have fancy big names such as hypoxylon canker or anthracnose or other technical names. All of these disease organisms hit trees for one reason and one reason only. When the health and the immune system of the plant is low, disease organisms enter the picture to do what nature set them up to do: come in and take out unfit plants. It is the same with insects, so the secret to controlling

Bare soil in the root zones of trees—something that should never be allowed. Photograph by Howard Garrett.

diseases of all kinds and insect pests is to increase the resistance by enhancing the immune system of the tree. That is done by working on the soil and using adapted plants. The specific treatment that we recommend is called the Sick Tree Treatment, and it is detailed in Appendix 1.

Oak Wilt and Other Tree Problems

Oak wilt is one of the most serious tree problems in Texas and many areas of the country. It is a devastating disease of native and introduced red oaks and live oaks. Texas A&M and the Texas Forest Service have been working on the problem for several years and recommend a program of trenching to separate the roots of sick trees from those of healthy trees, and a root flair injection of a chemical fungicide called Alamo. We don't recommend this program and have a different proposal that works well on most sick trees and shrubs.

To look at this problem from a little bit different angle, let's consider the insect that's blamed. The nitidulid or sap-feeding beetle, the alleged culprit, feeds on tree sap in the spring and spreads the disease. Adult beetles look like tiny june bugs. They inhabit the fungal mats beneath the bark of diseased red oaks (*Quercus texana* and *Quercus shumardii*). Infectious beetles emerge from the fungal mats and deposit oak wilt spores in wounds on healthy trees by feeding on sap. These are the same insects that feed on rotting fruit in the orchard.

There is research evidence now that sapsuckers attack and drill holes in trees that are in stress. The stress causes sweeter and more concentrated sugars. The sap beetles are probably attracted to trees in a similar way. But even if the little beetles go to both healthy and weak trees, why do some trees succumb to the disease while others do not? Even if the beetles don't infect every tree, the roots reach far out and touch the surrounding trees' roots. All the experts agree that the disease can easily spread through the roots. Even though the oak wilt disease has killed thousands of live oaks and red oaks in Texas, the disease can be stopped by using organic techniques. The plan is simple. Keep trees in a healthy condition so their immune system can resist the infection and disease. It has been noticed by many farmers and ranchers that the disease doesn't bother some trees—especially those that are mulched and those where the natural habitat under trees has been maintained.

Since the fungal mats form on red oaks only, not on live oaks, the live oak wood can be used for firewood without any worry of spreading the oak wilt disease. Red oak wood needs to be stacked in a sunny location and covered with clear plastic to form a greenhouse effect to kill the beetles and fungal mats. When oaks are shredded into mulch, the aeration kills the pathogens and eliminates the possibility of disease spread. That goes for all species.

Is this beetle the only way the problem could be spread? I doubt it. How about mechanical damage to tree trunks, wind, squirrels, hail, sapsuckers, and other insects? Fire ants seem to prefer weaker trees over others and could be part of the spreading problem.

Our recommended program has not yet been proven to work by any university and probably won't be, even though the evidence continues to stack up. Improving the health of the soil and thus the population of beneficial fungi on the root system seems to be paramount. Spraying the foliage during the rebuilding of the soil and root system provides trace minerals to the plant that can't yet come in through the roots. This program is not just for oak wilt. It works for most environmental tree problems and all tree types. My point here is that if it works for oak wilt, it will work even more effectively for less deadly tree conditions. If your tree problem is a result of poor variety selection, we can only help you in the future. Choose more wisely next time.

Nursery and Greenhouse Operations

The bed preparation in greenhouses varies depending on whether the plants are grown in the ground or in the pots. If the plants are grown in the ground, the typical bed preparation discussed under the Basic Organic Program works (see Appendix 1). If the plants are grown in pots, a potting soil needs to be used, so check the suggested formula at the end of Chapter 3 or use the one given below. One change would be the mulch that is used in the greenhouses. If the plants are planted in the ground, the mulch should be a cedar one laid over 2 inches of compost for optimum results. The cedar fragrance helps in repelling insects and diseases. It is also extremely helpful in preventing root knot nematodes. Cedar is far better than two flooring covers normally used: gravel and concrete. They are in fact the worst possible choices because these materials harbor pests and don't provide carbon dioxide as natural mulch does.

Potting soils for container growing should be compost based. Peat moss is expensive, dead, lacking in nutrients, and an environmental problem because of the wetlands destruction to mine or harvest it and the shipping needed to get the material to most of the country. Coconut fiber is a far superior product because it is light and fluffy but, unlike peat moss, stimulates biological activity. Peat moss's only good use is for storing bulbs and other perishable materials. Compost, on the other hand, is alive, loaded with nutrients, and inexpensive, and it recycles valuable local and regional natural resources and keeps those materials out of landfills. Various blends provide good results, but one mixture that has been successful is:

30% Compost
30% Coconut fiber
10% Decomposed granite
10% Zeolite
8% Greensand
7% Flora-Stim
5% Alfalfa meal

One of the first places we saw lava used as a soil amendment was in a greenhouse operation, specifically an orchid production. A grower in the Dallas area was using a combination of foam peanuts in the bottom of the containers with a 1–1½-inch layer of medium red lava gravel above that. He used no soil at all and he had the most beautiful orchids we have seen in a long time. When asked why he didn't use lava throughout the entire pot, the answer was that the lava rock held too much water in some cases for the orchids and it added considerable weight, which was a problem for shipping.

The most successful flooring for any greenhouse is cedar mulch, either fresh or de-oiled. Cedar can be put in anywhere, such as 2–4 inches thick on top of a drainage system to prevent supersaturation problems, but the fragrance from the cedar and the growth-enhancing characteristics are far superior to any other kind of flooring for greenhouses. Finely textured cedar flakes can also be used successfully to heal plants, as a soil amendment, and to control root knot nematodes. The organic (cedar or native) mulch on the floor of a greenhouse adds needed CO_2 to the air in greenhouses closed-up during winter. This stimulates plant growth, for plants have to have CO_2 to manufacture carbohydrates, which relate to stored energy.

An organic practitioner friend built a very unusual greenhouse on his farm. I used the basic idea and turned a storage building and a dog-run house into a greenhouse built specifically for winter propagation. The roof is solid (no skylights), and the only glass is on the south and east sides. During the summer, the only sun that enters the greenhouse is morning sun. In the winter, as the sun's angle lowers, more sunlight can enter the greenhouse. The angle of the roof is 22 degrees, which is the angle of the sun at the winter solstice, the lowest arc of the winter sun. During that time, the sun lights the entire interior of the greenhouse. Barrels full of water are heated during the day by the sunlight. At night they radiate warmth, keeping the heating costs to a minimum.

All greenhouse operations should incorporate a rainwater collection system for water and cost savings. The quality of the water will also be superior.

Recreational Properties

Golf Courses

Golf courses are a little harder to maintain organically because many golfers think that all golf courses should look like Augusta National Golf Club looks during the Masters Tournament every spring. Wanting to be like that golf course shrine is one of the biggest problems we have in getting more golf courses to use natural organic techniques. It has created an aversion to weeds of any kind. On the other hand, there is no reason why all golf courses can't be maintained organically, especially with regard to fertilizer, and still have extremely high-quality turf. A total organic program without weeds can exist—it simply takes more maintenance and a higher budget to do it. Wanting to be like Augusta National is fine—it's just that most courses don't have similarly unlimited budgets. The stark perfection of Augusta National is impressive, but it really doesn't make any sense for the average golf course. There are more costs to factor in than just the fertilizer and pest control products. The hidden costs to society exist in pollution, erosion, exposure of people to very toxic chemicals, health problems, and so on. What *does* make sense is to have more biodiverse turf, especially in the fairways and roughs, and to use a lot of wildflowers and native plants in out-of-play areas to reduce the overall costs, especially the cost of water. Water is the single largest cost in most golf course budgets. The reality is that most projects have budget constraints of

Tierra Verde Golf Club in Arlington, Texas—significant cost savings resulted from a switch to a 100 percent organic fertilization program. Photograph by Howard Garrett.

some kind. The organic program and its great water-savings advantage is a crucial point. The frequency of fertilizer and pest control application is also significantly reduced.

Golf courses, more so than parks and other landscapes, have been identified as major polluters. That perception (and reality) could easily be reversed if the golf course industry changed over to organic techniques. The problem is that people in the golf course industry follow the instructions disseminated by the local universities, the extension services, and the United States Golf Association (USGA). Superintendents are reluctant to strike out on their own and try techniques that aren't recommended by the establishment. Doing so requires that they take responsibility for themselves. When they use the chemical approach and it fails, all they have to say at that point is, "Well, I used the best available technology the industry has to offer—what the universities and USGA recommended—so it's not my fault." There is also comfort in doing what peers are doing. When the organic method is used, management has to take responsibility and make decisions based on common sense rather than on what someone wrote in a book or taught in a class many years ago. Most of the recommendations that are being used today were made by the chemical industry so that they can sell more of their products. One of the keys of the organic approach is that fewer inputs are needed the first year and are further reduced as time goes on. Once healthy, biologically active soil is established, the fertilizer input is greatly reduced and the pest control expenses are reduced to a fraction.

Some golf courses are starting to use a modified program. By adding compost, compost tea, volcanic rock powders, and beneficial organism products, superintendents are finding that they can cut back dramatically on the volume of very expensive synthetic high-nitrogen fertilizers. They also find that their pest problems diminish in a short time. For example, turf diseases are rare in an organic program.

Weaning the industry off the herbicides is the toughest part of the transition. At this writing, the only 100 percent organically maintained course in the United States is the Vineyard Golf Club in Martha's Vineyard, Massachusetts. The city golf courses in Arlington, Texas, have successfully installed a totally organic fertility program. Some synthetic herbicides are still used on a limited basis, and work to completely eliminate them is continuing. Tierra Verde Golf Club is the lead course of this group, and we have been given permission to use their current budget figures. To say that they are impressive is an understatement. They are shown below. This is a beautiful Audubon-certified (in the Audubon Cooperative Sanctuary Program) large championship-quality course. The overall budget is less than half of the budgets of high-end country clubs of lesser size in the Dallas/Fort Worth area. The most significant savings is in water use. According to Superintendent Mark Claburn (personal communication, November 16, 2007):

> The total acreage for the golf course is 257 acres—we only mow out 97 acres. Tierra Verde GC [Golf Club] totals for fertilizer in the past year on my greens [were] 6.19 N and 5.78 K per 1,000 sq. ft. per year. My totals for the golf course fertilizer were 1.5 # [lbs.] N in 2006 and 1.86 # [lbs.] N for 2007 (our year goes to October). We used the PPM 4–3–2 from Perdue Agrocycle (a chicken manure–based product).
>
> My botanical budget (fertilizers, soil amendments, and pest control products) is roughly $83,000.
>
> I will send precise amounts next week, but we spend about:
>
> $3,000 for compost
> $12,000 for top dressing and bunker sand
> $4,000 for other soil amendments
> $2,000 for turf paint (dormant greens)
> $21,000 for fertilizer for fairways, roughs, and tees (PPM and others)
> $3,800 on corn gluten meal

$5,000 on greens fertilizers and amendments
$3,800 on all chemicals used (almost all of this is [for] 1 pre-emergent application for crabgrass in the spring)

The rest is saved or used on sod and seed, and plants purchased (seasonal color, overseed).

Total annual budget is $640,000. We use an average of 63 million gallons of water per year. My preliminary information shows that most comparable courses in Texas use between 100 and 150 million [gallons] with annual budgets of $1.5–2 million.

What's important to note about the budget figures given above is that the water use and overall budget are substantially lower than those of country clubs and other public courses that have a similar quality. Another significant difference is the dramatically lower amount of nitrogen used. Mark has continued to maintain one of the three practice holes with synthetic fertilizers and discovered through tissue testing that organic fertilizer is 60 percent more efficient at getting into the plant than the artificial products are. The result is cost savings and a drastic reduction of nitrogen volatilization, leaching, and runoff. Besides being an environmental gem, this golf course is beautiful, and the quality of the turf is superb.

The other golf courses in Arlington—Meadowbrook Park, Lake Arlington, and Chester W. Ditto—are maintained with the same natural organic program.

Another Texas golf course using organic techniques is The Rawls Course at Texas Tech University in Lubbock, Texas. From the opening of the course in early 2010, Eric Johnson was the superintendent, and he used an organic fertilization program. His annual cost of pesticides is $3,000 compared to the average of $45,000 spent on other comparable courses. Eric estimated that he was saving about 300,000 gallons of water a day. The annual budget for the course was $700,000, and the turf was superb. Eric used some of the same techniques as Mark Claburn did in Arlington, but he also used a compost tea liquid fertilization program. Through the irrigation system, he injected liquid humate, molasses, seaweed, and liquid fish. He balanced the chemistry and biology of the soil with other natural organic techniques. The only negative to this story is that recently the course maintenance was contracted out to a large golf management company, and they had their own superintendent. It will be interesting to see how the quality of the course holds.

The only 100 percent organic golf course in the United States that we currently know about is the Vineyard Golf Club in Martha's Vineyard, Massachusetts. Jeff Carson is the superintendent. He and the club have won numerous environmental awards and have shown clearly that it can be done. It is completely organic because it had to be to get the permit to build the golf course. It is a beautiful, very high-quality golf course and has certainly set some important standards. The only issue with using this course as an example is that the budget is basically unlimited.

Golf Course Management Programs

There is no one foolproof program for golf courses. It depends on the soil, the grass types, the climate, and the budget. An example of an organic golf course management program that would work in general is as follows:

- Stop using the toxic products—fertilizers and pesticides.
- Stop using washed sand to top-dress greens and tees.
- Start top-dressing with a mix of fine-textured compost, greensand, zeolite, and/or lava sand. Zeolite alone can be used.
- Aeration holes should be filled completely when top-dressing.
- Stop fertilizing regularly with high-nitrogen or nitrogen-only fertilizers.
- Start using natural organic fertilizers.
- Start spraying or injecting into the irrigation system compost tea or a version of the Garrett Juice formula.

Note: There are indications that fertigation through the irrigation system with a mix of liquid humate, seaweed, molasses, apple cider vinegar, and liquid fish—in other words, Garrett Juice Plus—can replace a large portion if not all of the dry fertilizer requirements.

Special Needs Management

Sometimes the heavy use of chemicals and high-nitrogen salt-based fertilizers has created a completely unproductive soil. To change the situation in a short time frame and detoxify contaminated soil, the following program can be used.

- Aerate the contaminated area heavily.
- Apply high-quality compost at 1 ton per acre.
- Spray a mix of Garrett Juice or Garrett Juice Plus with orange oil at 2 ounces per gallon in a soil drench application. A one-time application at 20 gallons per acre is usually enough.

A second approach would be to drench the problem area with compost tea and a mycorrhizal fungi product such as Alpha Bio Thrive. Again, a one-time application at 20 gallons per acre is usually enough. As is the case with most of these approaches, increasing the rates of application is okay and even beneficial as long as the cost fits the budget.

The question comes up often about what's better—compost or compost tea. Research has shown that it takes 5 applications (5 trips) of compost tea at a concentration of 20 gallons per acre to equal just one application of only one ton (1.5 cu. yd.) of compost, hence compost is far more economical. Two applications of tea gave minimal results, whereas 1 cubic yard of compost thinly spread made the grass start growing right away. Compost tea is a valuable tool in our toolbox, but it has possibly been overhyped and is not as cost-effective as other product combinations and techniques.

Alternate Maintenance Program

Another approach for successful golf course management could go something like this.

Fertilizer

First year: The program here can be very flexible. One transition program that works well is to apply humate or micronized compost at 100–200 lbs. per acre, urea at 100 lbs. per acre, dry molasses at 200–300 lbs. per acre—three times per year (February, May, September). For a pure organic approach, the urea can be replaced by doubling the rate of micronized compost.

After first year: Apply natural organic fertilizer 2–3 times per year at 20 lbs. per 1,000 square feet in early and late spring and in fall. Supplement with rock powders such as greensand, lava sand, soft rock phosphate, and/or zeolite. Rates will vary depending on soil chemistry and biological activity.

Compost

- Apply ½ inch to weak turf areas. Begin on-site composting from collected grass clippings, food waste, landscape waste, etc. Use this compost primarily for the landscaped areas.
- Use manufactured compost products for greens, tees, and fairways because of better consistency.
- See Chapter 6 for more details.

Aeration

- Core-aerate in spring and fall for the first three years. Leave holes open on tees and fairways.
- Fill holes on greens and tees with a mixture of compost, lava sand, and greensand and spread ¼-inch layer of the mix on top.
- Many golf courses aerate greens too often and too many times

during the year. When using organic products and techniques, the biological activity of the soil will take care of the aeration naturally.

Top-dressing Divots on Fairways and Greens

- Fill divots on tees and fairways with a compost, lava sand, and greensand mix. Do not use green-dyed sand. It's expensive and looks terrible. Straight greensand can be used if the mixing is a problem.
- For a less expensive mixture, use 50 percent washed sand, 50 percent fine-screened compost, and 5 pounds of greensand per cubic yard.

Foliar Feeding

- All turf areas will benefit from regular foliar feeding if the budget allows.
- Use compost tea or Garrett Juice at 20 gallons per acre.
- Use mycorrhizal products through fertilization systems or in sprays that are applied to the ground. Biological products can be omitted when spraying tree and shrub foliage. There would be no damage, but many of the microbes won't be activated to help until the product reaches the soil.

Weed Control

- One option is to control winter weeds by applying corn gluten meal in early spring, early summer, and early fall (e.g., 3/1, 6/1, and 10/1—if not overseeding in fall—in North Texas) at 20 lbs. per 1,000 square feet as preemergent weed prevention. This is a technique that is being used less and less. It's expensive and the timing is tricky to figure out.
- Spot-pull troublesome weeds or spray with 20 percent vinegar plus ½ cup of orange oil per gallon during full sun. Use mechanical weeders such as the Weeder, the Weed Hound, and the Weed Popper.

Insect Control

- Release beneficial insects for first two to three years or until natural populations establish. Trichogramma Wasps: Release 10,000 eggs per acre weekly starting at leaf emergence and continuing for three months.
- Green Lacewings: In April, release 2,000–4,000 eggs per acre weekly.
- May–September: Release 1,000 eggs of green lacewings per acre every two weeks.

- Apply beneficial nematodes in fall of first year and spring of second to control grubs and fire ants. Drench problem mounds with citrus-based products.

Disease Control

- For fungal diseases, apply horticultural cornmeal at 20 pounds per 1,000 square feet. Then, apply only as needed to control fungal diseases and molds at 20 pounds per 1,000 square feet. Apply mycorrhizal fungi products to problem areas per label directions.
- Spray Garrett Juice plus potassium bicarbonate to spot-treat small problem areas. GreenCure is a new commercial product. Spray Bio Wash for most fungal diseases, spray 3 percent hydrogen peroxide for bacterial diseases, and apply corn gluten meal for dollar spot.
- Landscape perimeter with native shrubs, trees, and grasses to create a biodiverse landscape and wildlife habitat.
- Mulch around trees and shrubs with shredded native tree trimmings (native mulch).
- Apply Sick Tree Treatment to sick trees where necessary.

General

Growth regulators such as Primo should not be used on turf. The theory is that these products save money by reducing the number of mowings. These products change the grasses so that they grow more laterally than upright. The result is horizontal, grainy turf that creates very inconsistent playability. A better plan is just to mow more often. On the other hand, the organic program results in less mowing.

Parks

Parks are easy, especially compared to golf courses. No park in the world should ever have to use a single toxic chemical. Parks do not need the high-quality, intense, expensive maintenance that golf courses require for greens and tees. Parks should be clean, pleasant places with healthy soil, native plants, and no toxic chemicals. One major fertilizer application a year along with a rock powder application is generally all that is needed. Parks that have a considerable amount of compaction from foot or vehicular traffic can benefit from mechanical aeration or the application of one of the biological products such as Bioinoculant, Agri-Gro, or Thrive. Compost applications are greatly helpful if the cost can be fit into the budget. Natural organic fertilizers and mulches on the bare soil are sometimes enough. Additional organic inputs are certainly helpful if the budget allows.

Using mulches heavily can be one of the most cost-effective tools for parks. Native shredded mulches are inexpensive for the benefits they provide as they prevent erosion, build the soil, and slowly feed plants, eliminating the need to buy more expensive fertilizers. This is a particularly effective approach when lots of trees exist.

Trees like mulch as a groundcover under them much better than grass or another groundcover, plus mulch is cheaper and it looks good. This mulching technique really works well for parks but is also appropriate for golf courses in out-of-play areas. Even in rough areas that do come into play, a double-ground or finer-textured shredded mulch works well. You just don't want large pieces flying up to hit golfers in the face. Some parks and golf courses use pine needles as mulch. That only looks right under pine trees, and it can catch on fire and smolder if someone throws a cigarette into it.

For walkways, use decayed granite, basalt gravel, or other mulch. Use the least amount of concrete and other pavements that would require cutting or excavation around trees. Tree pruning in parks should be kept to a minimum, removing primarily any limbs that could cause a dangerous condition. Apply Tree Trunk Goop to trunk and limb wounds on trees where necessary.

Sports Fields

Sports fields are a little bit harder to maintain organically but not by much. Use the Basic Organic Program but reduce the applications as necessary to meet the budget. Compost as a mulch on turf can be used on large commercial properties and parks and athletic fields as well. High school football fields that have been composted have been reported to have fewer injuries, especially knee injuries and shin splints, due to the healthy, soft, thick turf on softer soil. Even with heavy play, the healthy soil resulting from the organic program rarely has the hard compaction issues that result from the synthetic management approach. Often the medical savings from not treating so many injuries pays for the whole program.

Not all composted sports turfs have reported championships, but they are all reporting less need for fertilizers, irrigation, pesticides, and herbicides as the thick turf chokes out the weeds. Additionally, all compost users are reporting less need for irrigation, with 30–50 percent less water needed to maintain the turf. Mulching the lawn with compost in the fall is the closest thing to a cure-all there is. Although it is work spreading the compost, we know of no one who has been disappointed. A range of equipment to apply compost to turfgrass now exists, from trucks with blowers to Power Take-Off–driven spreaders to small walk-behind spreaders from many companies. The application can be contracted out or one can rent the required equipment.

Test spot treated with compost on a sports field in Houston, Texas. Photograph by John Ferguson.

Some sports field managers have reported success from programs as simple as one annual application of dry molasses at 200 pounds per acre.

The simplest plan for sports fields is to add compost, rock powders, and molasses. Leave the clippings on the ground and cover all bare soil with mulch. As with most projects, all sports fields will greatly benefit from compost products that are laced with mycorrhizal fungi.

Best of all, there is no need to expose sensitive children and youth to toxic chemicals that hurt their health and cause hormone disruption.

CHAPTER 10

Organic Strategies and Global Climate Change

Why should landscape architects and contractors, commercial growers, land managers, and other stewards of the land be concerned about global climate change? Global warming can potentially cause several significant problems for farms, ranches, nurseries, and landscapes: extreme and unpredictable weather, coastal flooding, regional droughts, species extinctions, ecosystem disruptions, and reduced pollination of many important plant species. These changes, in combination with dwindling fossil fuel supplies and deteriorating water quality, will affect the ability of our lands to support a growing global population and our ability to provide stable landscapes.

The Intergovernmental Panel on Climate Change (IPCC) has found global warming to be "unequivocal" and states with unprecedented certainty that this warming is due to greenhouse gas emissions. Indeed, the dependence of growing operations on fossil-fuel-derived products is alarming and has created significant pollution problems for our atmosphere, rivers, lakes, groundwater, and oceans. These threats are indeed serious, and many growers, including nursery, turf, and land managers, around the world believe they are witnessing such effects already. Professionals are concerned and asking the question: What impact do farm, nursery, turf, and landscape management practices have on global climate change?

Global Warming by the Numbers

Here are some of the scientific observations that are driving efforts to look for ways to ameliorate the effects of global warming. Atmospheric carbon dioxide (CO_2) levels have risen from 280 parts per million (ppm) to 365 ppm in the last two centuries. The current rate of increase is 1.3 ppm per year. Atmospheric CO_2, the major factor in the greenhouse effect, is now twice as high as it was during the last Ice Age. Consumption of remaining fossil fuel

Satellite image of damage to Gulf of Mexico caused by runoff of synthetic fertilizers from agricultural crops. Photograph courtesy of USDA.

reserves would boost CO_2 by a factor of four to eight. Paleoclimatic evidence indicates that the 1990s were the warmest decade since the year AD 1000.

Scientists are busy developing an increasingly more accurate picture of the global carbon cycle in their efforts to understand the effects of global warming. Total carbon storage and annual fluxes have been established for various parts of the global carbon system: terrestrial vegetation, the ocean surface, the deep ocean. Carbon dioxide emissions from human and animal activities are estimated at about 8.9 billion tons per year, while net atmospheric CO_2 accumulation is 3.5 billion tons. Thus, 40 percent of annual human-induced carbon emissions contribute to increases in the atmosphere, while the remaining 60 percent are absorbed by terrestrial plants and oceans.

Where can we put all of this extra carbon? Significant quantities can be put back in the ground. Provocative new data are emerging that implementing organic methods can immediately begin to reduce global warming emissions. Simply put, it's time to change how we raise our food and manage our landscapes. For years, many have argued that organically produced food is safer and more nutritious. Now we are learning that a switch to organic methods is an expedient and soil-based sink for carbon from the atmosphere. Data from the Rodale Institute's long-term comparison of organic and con-

Organic trials at the USDA fields in the Rio Grande Valley in Texas. These cotton plants are being grown under the traditional chemical program. Photograph by Howard Garrett.

Cotton plants showing greater growth and more bolls per plant in the fields in Joe Bradford's organic trials. This field uses organic fertilization and no toxic pesticides. Photograph by Howard Garrett.

ventional farming methods substantiate that organic practices are much more effective at removing carbon dioxide, the greenhouse gas, from the atmosphere and fixing it as beneficial organic matter in the soil. Organic practices result in rapid carbon buildup in the soil. This organic approach does not rely on high-tech or space-age solutions; instead it takes advantage of the symbiotic relationships between beneficial soil organisms that have been successful at maintaining the productivity of the land for hundreds of millions of years.

On a global scale, soils hold more than twice as much carbon, an estimated 1.74 trillion tons, than does terrestrial vegetation (trees, shrubs, and grasses), which holds 672 billion tons. Research is revealing that practices such as the use of cover crops, compost additions, and biological inoculants can dramatically alter the carbon storage of managed lands. Switching from conventional to organic methods switches land management from a major global warming problem to a major global warming solution. Managing farms, landscapes, trees, and turf organically also provides a sink for atmospheric carbon. Cover crops, organic fertilizers, compost, and mulch added to farms, nurseries, and landscapes support the activities of beneficial soil organisms and the buildup of carbon in the soil compared to intensive conventional management, which depends upon fossil-fuel-derived fertilizers and pesticides and can reduce carbon in the soil.

The Rodale Institute and researchers at Cornell University have collaborated to develop some extraordinary numbers related to organic growing operations in the United States and carbon cycling. U.S. agriculture currently releases 1.5 trillion pounds of CO_2 annually into the atmosphere. Converting all U.S. agricultural lands to organic production would eliminate agriculture's massive emission problem. Switching to organic production would actually result in a net increase in soil carbon of 734 billion pounds.

If just 10,000 medium-sized farms in the United States converted to organic production, they would store so much carbon in the soil that it would be equivalent to taking 1,174,400 cars off the road or reducing car miles driven by 14.62 billion miles!

Organic Methods Make a Difference

The Rodale Institute Farming Systems Trial (FST) compared conventional and organic methods for twenty-three years in side-by-side trials. This research is a wealth of information about the ecological and economic benefits of organic growing and includes detailed studies of mycorrhizal fungi, weeds, compost, cover crops, water quality, and profitability. It contains convincing evidence that growing organically helps combat global warming by capturing atmospheric carbon dioxide and incorporating it into the soil, whereas conventional practices exacerbate the greenhouse effect by producing a net increase of carbon in the atmosphere.

The key to this process lies in the handling of soil organic matter, or SOM. Because soil organic matter is primarily carbon, increases in SOM levels directly correlate with carbon sequestration. While conventional farming practices such as plowing and leaving soil bare typically deplete SOM, organic farming builds it through the use of composted animal manures, cover crops, legumes, and the activities of symbiotic soil fungi and bacteria. In relationship to global warming, it may be that the soil itself and the life that resides in it can make a major difference. "Good organic growers see carbon as a resource that improves crop quality, soil productivity, and yield rather than carbon as a waste product," says Gerald Wiebe, an organic farmer in Manitoba, Canada. Gerald sees the value of active management of the biological component in soil to encourage carbon storage in soil. "Some organic growers tend to use a zero input approach to organic farming, thinking that sunshine, seed, and rain are all they need. This often leads to decreased yields with no improvement in soil conditions. The simple nonuse of conventional inputs does not guarantee that soil microbial life, carbon levels, and soil structure will be restored. However, when mycorrhizal and bacterial inoculants, compost, compost tea, green manures, legumes, etc. are used in a well managed program, positive soil changes happen relatively quickly."

Launched in 1981, the FST is a long-term experiment comparing three agricultural management systems: one conventional, one legume-based organic, and one manure-based organic. In twenty-three years of continuous recordkeeping, the FST's two organic systems have shown a 15–28 percent increase in soil carbon, while the conventional system has shown no statistically significant increase. For the organic systems, that converts to more than

1,000 pounds of captured carbon (or about 3,670 pounds of carbon dioxide) per acre-foot per year. These statistics improve when considering the reductions in carbon dioxide emissions represented by the organic systems' reduced use of fossil fuels. A comparative analysis of FST energy inputs, conducted by Cornell University, found that organic farming systems use just 63 percent of the energy required by conventional systems, largely because of the massive amounts of energy required to synthesize nitrogen fertilizer in conventional systems.

So just how much carbon dioxide can growing organically take out of the air each year? As mentioned, organic farms sequester as much as 3,670 pounds of carbon dioxide per acre-foot each year. A typical passenger car traveling an average of 12,500 miles per year, according to the EPA, emits 10,000 pounds of carbon dioxide a year. One way to imagine the benefits is to think of the equivalent number of cars that would be taken off the road each year by growers converting to organic production. For example, if all 149 million acres of conventionally grown corn and soybean fields in the United States were converted to organic production, that could translate to:

- 54.8 million cars off the road
- That's 685 billion car miles not driven . . . or 116,666,666 round-trips from New York City to Los Angeles not taken!

If all 442 million acres of U.S. cropland were converted to organic, that would be equivalent to:

- 162 million cars being taken off the road (over half of the national total)
- 2 trillion car miles not driven.

The Link to Soil Life

The land-based carbon cycle works like this. Carbon dioxide is taken out of the atmosphere by plants and converted to organic material through photosynthesis. The oxygen in the molecule is released back into the air, and the carbon becomes part of the plant structure and is pumped down into the soil to fuel the activities of beneficial soil organisms. The carbon that is absorbed from the atmosphere by plants and animals can take several paths before it reenters the air as carbon dioxide. When a plant or animal dies, it is broken down by soil microorganisms that feed on the dead organic matter. As the microorganisms consume the organic matter, they release some of the carbon into the atmosphere in the form of carbon dioxide.

Some carbon is destined for longer-term storage in roots and in the bodies of plant-eating or carnivorous animals. Finally, as plants and animals decay, instead of escaping as carbon dioxide, a significant portion of their carbon becomes part of the organic component in soils through the activities of essential soil organisms. These beneficial microorganisms work to produce a substance known as humus, a stable, rich component of soil that is the color of dark chocolate, and glomalin, a sticky organic glue, both of which are loaded with carbon.

The work of another Rodale research collaborator, David Douds of the USDA Agricultural Research Service, suggests that one important group of beneficial soil organisms, the mycorrhizal fungi, deposit carbon in soil. In the FST, soils farmed with organic systems have much greater populations of mycorrhizal fungi. Overwintering cover crops supply energy to fuel the activities of mycorrhizal fungi in the organic system, in contrast to the conventional systems that have a significantly greater fallow period. Cover crops allow plants to occupy the land in all seasons and sustain the energy requirements of beneficial mycorrhizal fungi. Thus plants can capture carbon dioxide year-round and protect the soil from erosion and nutrient runoff.

Other benefits to the proliferation of the mycorrhizal fungi have been profound. In another three-year study of mycorrhizal fungi at the Rodale Institute, pepper and potato yields increased 34 percent and 50 percent respectively compared to controls. Douds's research suggests that a small amount of mycorrhizal fungi can be substituted for a large amount of fossil fuel-gobbling chemical fertilizer in the growing of crops.

Utilizing the mycorrhizal relationship has global implications for the strategy to combat global warming. Most plants, including more that 90 percent of all agricultural crops, form a root association with these specialized fungi when they are present. Mycorrhizae literally means "fungus roots" and represents a symbiotic association between fungus and plant. Fungal filaments extend into the soil and help the plant by gathering water and nutrients and transporting these materials back to the roots. In exchange, the plant supplies sugar and other compounds to fuel the activities of the fungus. Miles of fungal filaments can be present in an ounce of healthy soil. The crop's association with mycorrhizal fungi increases the effective surface-absorbing area of roots several hundred to several thousand times. This is an example of symbiosis, a win-win association.

Carbon-rich Organic Glue

Mycorrhizal fungi are also performing another soil carbon investment service that has only recently come to light. In 2002, the journal *Plant and Soil*

Corn-growing research by Mike Amaranthus. Tote on the left used organic fertilizer and mycorrhizal fungi. Tote on the right used recommended synthetic fertilizer. Photograph by Mike Amaranthus.

published a report by M. C. Rillig, S. F. Wright, and V. T. Eviner ("The Role of Arbuscular Mycorrhizal Fungi and Glomalin in Soil Aggregation") that suggests that the substance glomalin, discovered by Wright in 1996, is a mechanism for storing large amounts of carbon in soil. Glomalin, produced by the mycorrhizal fungal group Glomales, hence the name glomalin, is produced by mycorrhizal fungi established on a plant's roots. An organic "glue," the glomalin molecule is made up of 30–40 percent carbon and represents up to an astonishing 40 percent of the carbon in soil. Since soils globally contain 1.74 trillion tons of carbon, a 40 percent contribution of soil carbon by mycorrhizal fungi equals 696 billion tons, or more than all of the terrestrial vegetation (672 billion tons) on earth!

Glomalin acts to bind organic matter to mineral particles in soil. It also forms soil clumps, aggregates that improve soil structure and deposit carbon on the surface of soil particles. It is glomalin that gives soil its tilth, a subtle texture that enables experienced farmers to identify rich soil by feeling for the smooth granules as they flow through their fingers. The mycorrhizal fungi apparently produce glomalin to seal themselves and gain enough rigidity to carry materials across the air spaces between soil particles. Wright's discovery of glomalin is causing a complete reexamination of what makes up soil organic matter. It is increasingly being included in studies of carbon storage and soil quality. Jim Trappe, USDA mycorrhizal researcher and professor emeritus at Oregon State University, says, "Adding carbon to soils is not just an inert chemical process, rather it is profoundly influenced by the biological activity in the soil. Organic methods favor the abundance and diversity of soil life."

Organic Material, Organic Matter, and Soil Carbon

Many times we think of organic matter as the plant and animal residues we incorporate into the soil. We see a pile of leaves, manure, or plant parts and think, "Wow! I'm adding a lot of organic matter to the soil." This soil amendment is actually organic material, not organic matter.

What's the difference between organic material and organic matter? Organic material is anything that was alive and is now in or on the soil. For it to become organic matter, it must be decomposed into humus. Humus is organic material that has been converted by microorganisms to a resistant state of decomposition. Humus is approximately 50 percent carbon and 5 percent nitrogen.

Organic matter is stable in the soil. It has been decomposed until it is resistant to further decomposition. Usually, only about 5 percent of it mineralizes yearly. That rate increases if temperature, oxygen, and moisture conditions become favorable for decomposition, which often occurs with soil tillage. It is the stable organic matter that is analyzed in the soil test. Organic material, on the other hand, is unstable in the soil, changing form and mass readily as it decomposes. As much as 90 percent of it disappears quickly because of decomposition.

How much organic matter and carbon is in the soil? An acre of soil measured to a depth of one foot weighs approximately 4,000,000 pounds, which means that 1 percent organic matter in the soil would weigh about 40,000 pounds per acre and contain 20,000 pounds of carbon. Since it takes at least 10 pounds of organic material to decompose to 1 pound of organic matter, it takes at least 400,000 pounds (200 tons) of organic material applied or returned to the soil to add 1 percent stable organic matter (40,000 pounds) under favorable conditions.

Squashing the Symbiosis

Unfortunately, many conventional growing, turf, and landscape management practices reduce or eliminate mycorrhizal activity in soil. Certain pesticides, chemical fertilizers, intensive cultivation, lands lying fallow, soil compaction, organic matter loss, removal of natural vegetation, and erosion adversely affect beneficial mycorrhizal fungi. An extensive body of laboratory testing indicates that the majority of intensively managed lands lack adequate populations of mycorrhizal fungi. High levels of chemical fertilizers not only have a high energy cost but also have a devastating effect on life in the soil. Chemical fertilizers are basically salts, which means they suck the water out of beneficial bacteria, fungi, and a wide array of other organisms in the soil. Arden Anderson, both a PhD agronomist and a practicing physician, states that "conventional agricultural practices with an emphasis on pumping the soil with chemical fertilizers can destroy soil life, which in turn affects the fertility of the land and nutritional value of our food."

Research has shown that these beneficial soil organisms form the basis of the food web, which conserves and processes nutrient capital in the soil. Without this soil food web, a substantial amount of nutrients are leached from soil into waterways, where they damage water quality and aquatic life. Many conventional practices impact large segments of living soils where a large quantity of nutrient capital and soil organic matter is lost, and the grower and landscape professional is forced to add more fertilizer. The job once accomplished by beneficial soil organisms must then be done by the grower or landscape manager.

Managing Trees in Landscapes as Carbon Sinks

Trees (and shrubs) are unique among plants in that they have a woody stem and roots that get bigger every year, and these woody parts last for decades or even centuries. Since this wood is mainly made of carbon from the greenhouse gas carbon dioxide, tree stems and roots are good long-term storage places for carbon. Trees are, in reality, solidified lumps (beautiful nonetheless) of carbon that have been created by drawing carbon dioxide out of the atmosphere.

Carbon sequestration rates vary by tree species, soil type, regional climate, topography, and management practice. Soil carbon sequestration rates for trees depend on tree height, age, and growth rate. In the United States, fairly well-established values for carbon sequestration rates are available for most tree species ("Calculating Carbon Sequestration by Trees," U.S. Dept. of Energy, 1998).

For example, let's look at an example of a landscape project in the United States that would involve the planting of one hundred balled and burlapped nursery-raised trees (50 Douglas fir and 50 Norway maple) of about 1-inch caliper. In year one, the annual sequestration rate for the Norway maples would be 2.7 pounds of carbon per tree and for the Douglas firs, 2.2 pounds of carbon per tree. At year fifty, each moderately growing maple is annually sequestering 67.2 pounds of carbon and the fast-growing Douglas fir 106.3 pounds of carbon. Over the fifty-year period, the one hundred trees have sequestered over 185,000 pounds of carbon, or 680,000 pounds (340 tons) of carbon dioxide, from the atmosphere. Planting 50 million trees in urban areas next year would sequester approximately 170 million tons of carbon dioxide in the next fifty years. Landscape professionals appreciate that trees growing in the suburban and urban landscape have a great opportunity to draw carbon dioxide out of the atmosphere. Using quality compost and native mulch in planting and maintaining trees not only directly adds carbon to the soil but also encourages increased root growth, which is another important carbon sink.

Trees can also play a role in helping reduce greenhouse gas emissions mainly by reducing the use of fossil fuels (coal, oil, and gas). In the hottest part of the summer, about half of the electricity used in the United States powers air conditioners, and air conditioning causes power plant emissions of 100 million tons of carbon each year. Trees planted to shade buildings and cool the air through transpiration can reduce this energy use by up to 70 percent. Well-placed trees that slow the wind can reduce energy use for heating by 30 percent. Trees in living snow fences reduce the energy needed to plow roads and parking lots. These are just some of the ways that trees can be used to save energy, thereby reducing fossil fuel use and carbon dioxide emissions.

The simple act of mulching around the tree can help conserve water, moderate soil temperatures, and add a significant amount of carbon to the soil. As the mulch breaks down, it also provides a source of nutrients for the tree, reducing the need for fertilizers that not only contribute to water pollution but often take significant amounts of fossil fuels to produce.

Nursery managers, arborists, and landscape professionals across the country are promoting appropriate tree planting, both urban and rural, because of these many benefits that trees provide. Trees can play an important part in reducing fossil fuel consumption and carbon dioxide emissions as well as sequestering carbon dioxide out of the atmosphere. Few global-warming-fighting tools provide such diverse benefits at such a low cost for such a long period of time.

Landscape and Turf Management

Landscape and turf professionals are now providing homeowners and municipalities with information to help them manage their landscape in an ecologically sound manner not only because it is best for the environment but also because many environmentally sound practices result in healthier and more productive landscapes. One of the objectives of this book is to inspire growers and turf and landscape professionals to minimize their contributions to climate change and create a better environment.

There's a perception among some people that "doing the right thing" to reduce global warming requires sacrifice and extra work, but when it comes to growing and managing landscapes, the reverse is often true. For example, replacing large expanses of intensively managed turfgrass with low-maintenance grasses or native plants reduces the time and expense of mowing, fertilizing, and watering. Taking steps to conserve and protect water resources, using organic fertilizers, and using biological inoculants are all elements of "ecological" turf and landscape management that reduce our greenhouse gas emissions and may, when combined with compost and mulch additions, actually store significant amounts of carbon in soil. Minimizing the use of gasoline-powered equipment is also vital. However, another often overlooked technique involves using plants for energy conservation. In the era before people could flip a switch to turn on the heat or air conditioning, landscaping played an important role in the comfort of a home. People planted trees and shrubs in strategic locations to mitigate hot summer sun and cold winter winds while making the most of cooling summer breezes and radiant heat from the winter sun.

Turf professionals are discovering how to grow beautiful lawns, sports fields, and golf courses that are more eco-friendly. Avoid chemical fertilizers, the performance-enhancing drugs of lawn care. While the blast of nutrients spurs rapid growth and a deeper green color, it comes at a great cost. In addition, the extra, unabsorbed nitrogen and phosphorus can contaminate surface and groundwater, kill beneficial microorganisms, attract the wrong kinds of insects, and promote turf diseases. Organic fertilizers and quality compost provide a wider range of nutrients that come from sources that decompose and release slowly. In some cases, switching to a more environmentally friendly turfgrass species is the best choice. Some turfgrasses, such as Zoysia, Paspalum, Buffalo, and Centipede, require significantly less fertilizer and water inputs, lower maintenance, and infrequent mowing. Organic turf areas are just as beautiful, safer, sustainable, and able to defend themselves from insects and disease.

Turf and landscape professionals are reinvigorating their programs to incorporate organic methods and common sense. Perhaps our movement away from these in the past was simply due to expediency: after all, trees get in the way during construction. Or perhaps it had to do with our culture's obsession with the perfect lawn. No doubt it had to do with our legacy of cheap electricity and fuel. When electricity was inexpensive and heating oil was fifty cents a gallon, conserving these resources wasn't a high priority. But it's not too late to turn things around. Professional growers and turf and landscape managers can make a real difference.

Landscape and turf professionals who use organic fertilizers, mulch, and compost directly increase the carbon content of soil. The widespread use of quality compost and mulch accomplishes more than simply adding carbon to the earth's soil. Compost and mulch also conserve water. A native mulch layer of leaves, twigs, grass, compost, or any organic material from humans' waste stream will protect the soil from the baking sun and drying winds. Carbon-enriched soil can absorb and hold more water. The mulch holds heavy rains in place until they soak in. This prevents floods and soil erosion, which also carry soil organic matter from the land. Growers and landscape and turf professionals as well as homeowners can benefit from the use of compost and mulches on their landscapes. Compost applied in the fall and watered in well will do more to keep a lawn and landscape healthy than the best chemical program. Compost acts as a chelating agent, preventing macronutrients, such as phosphorus, and micronutrients, especially zinc and iron, from locking up in alkaline soils.

Conclusions

While the world has been focused on reducing carbon emissions and water pollution from industry it has forgotten about one of the largest contributors to the greenhouse gas and nutrient runoff problem: our synthetic approach to food production, nursery production, and turf and landscape management. The way we grow our food today and manage the land has contributed in a major way to our environmental problems. The good news is that a change to organic methods could be a powerful solution to global warming and water pollution problems.

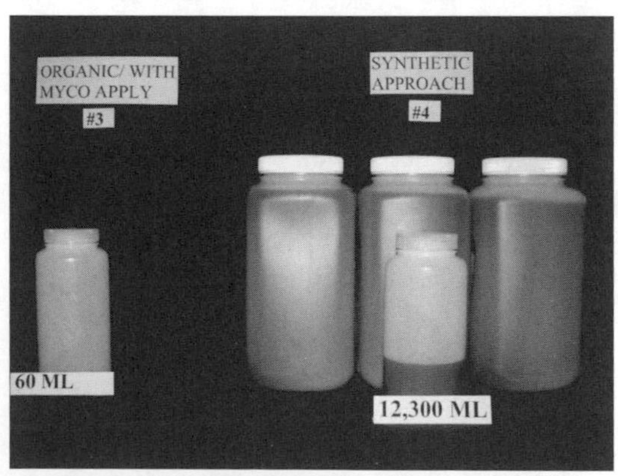

Containers showing the dramatic reduction of water runoff and lost nutrients in the organically fertilized plants (left) compared with the synthetically fertilized plants (right). Photograph by Mike Amaranthus.

Organic programs implemented on a large scale can make a difference in the quality of our environment, our use of fossil fuels, and our climate. Whether it is related to carbon, water, fertilizer, or fossil fuels, the organic approach applied on a large scale has enormous potential to create a more sustainable future for everyone. As stewards of the land, growers and turf and landscape professionals can make a difference. "It's time to get going," says Oregon State University Professor Dave Perry. "Implementing the organic approach is a tried and true method that not only takes carbon out of the atmosphere but provides an abundance of additional environmental benefits."

We invite you and other professionals to join our effort to leave the legacy of a healthy planet for generations to come. After all, we are all in this together.

Appendix 1. Organic Treatment Formulas

Basic Organic Program

Stimulating and maintaining healthy biological soil is the key. It's not complicated—simply avoid doing anything that hurts the life in the soil and choose only those inputs that benefit the life in the soil and that make sense from a horticultural and economic standpoint.

1. **Stop** using all synthetic fertilizers, toxic pesticides, and other chemicals that harm living organisms. All high-nitrogen synthetic fertilizers are bad, and nitrogen-only products are the worst.
2. **Build** soil health with aeration, compost, rock minerals, sugars, and microorganism products.
3. **Use** native plants and well-adapted introductions, water carefully, mulch bare soil, and monitor plant growth.

Planting

Bed preparation: Scrape away existing grass and weeds; add compost, lava sand, organic fertilizer, cornmeal, dry molasses, and rock minerals and till into the native soil. Excavation of natural soil and additional ingredients such as concrete sand, peat moss, foreign soil, and pine bark should not be used. More compost is needed for food crops, shrubs, and flowers than for turf and groundcover. Add greensand to alkaline soils and high-calcium lime to acid soils. Decomposed granite, rock phosphate, and zeolite are effective for most all soils.

Maintenance

Fertilizing: Broadcast organic fertilizer over the entire site one to two times per year at 20 pounds per 1,000 square feet. Foliar-feed all plants during the growing season at least monthly with liquid humate or Garrett Juice Plus. High-nitrogen salt fertilizers and products that contain synthetic material must be eliminated. Sewage sludge, biosolids, Miracle-Gro, Peters, other soluble crystal-type products, and Osmocote are also not acceptable in an organic program.

Mulching: Mulch bare soil around all shrubs, trees, groundcovers, and food crops with shredded native tree trimmings to protect the soil from sunlight, wind, and rain; inhibit weed germination; decrease watering needs; and mediate soil temperature. Other natural mulches can be used, but avoid Bermuda grass hay because of herbicide residue. Also avoid pine bark, cypress mulch, rubber products, and chemically dyed wood products. Do not pile mulch on the stems of plants.

Watering: Water only as needed. The organic program will reduce the frequency and volume of water needed. Add a tablespoon of apple cider vinegar per gallon of water when watering pots. Use 1 ounce of liquid humate in acid soils. Garrett Juice can be used in either case. Be careful of drip irrigation systems because with those systems, it is difficult to avoid dry and water-logged spots. Watering from above with sprinklers is usually best. It rains from above.

Mowing: Mow turf as needed and mulch clippings into the lawn to return nutrients and organic matter to the soil. Put occasional excess clippings in the compost pile. Don't ever let clippings leave the site. Do not use line trimmers around shrubs and trees. Buffalo grass lawns need less care than those of other grasses.

Weeding: Hand-pull large weeds and work on soil health for overall control. Mulch bare soil in beds. Avoid all synthetic herbicides, including Roundup, 2,4–D, MSMA, preemergents, broad-leaf treatments, soil sterilants, and especially the SU (sulfonylurea) herbicides such as Manage and Oust. Spray noxious weeds as needed with vinegar-based or fatty-acid herbicides.

Pruning: Do not "lift" or "gut" trees. Remove dead, diseased, and conflicting limbs. Do not overprune. Do not make flush cuts. Leave the branch collars intact. Do not paint cuts. All of this is artificial and hurts trees.

Controlling insect pests: In general, control insect pests by encouraging beneficial insects and microbes and spraying with the Garrett Juice Plus mixture. Spray minor outbreaks with plant-oil products, including orange oil, garlic-pepper tea, and essential oils. Avoid all pyrethrum products, especially those containing piperonyl butoxide (PBO), petroleum distillates, and other contaminants.

Controlling diseases: Most diseases such as black spot, brown patch, powdery mildew, and other fungal problems are controlled by prevention through soil improvement, avoidance of high-nitrogen fertilizers, and proper watering. Outbreaks can be stopped with sprays of potassium bicarbonate, cornmeal juice, diluted milk, or the commercial product Bio Wash. Bacteria and viruses are controlled with 3 percent hydrogen peroxide, but it can burn plants when sprayed during the hot part of the day.

Soil amending: Apply compost; rock materials such as lava sand, granite, basalt, or zeolite; and dry molasses to all planting areas. Use products that introduce and/or stimulate beneficial microbes in the soil.

Treating sick or weak plants: Drench the root zone with Garrett Juice Plus and New Plant THRIVE or other quality microbe products. Apply the entire Sick Tree Treatment (see formula below) for best results.

Compost Tea

To make quality compost tea, start with a well-made fungus-filled compost that contains aerobic beneficial microbes, then multiply them by feeding and aerating with a simple aquarium pump to increase the number and variety of microbes, including bacteria, fungi, protozoa, flagellates, and beneficial nematodes. Typical garden soils are weakest in fungal species, and this procedure helps them greatly. Buy a pump rated for about a 50-gallon aquarium. Do not overdo the movement of tea and beat fungus to death, but do provide enough oxygen to keep the tea from going anaerobic and smelling.

Air stones, also from the pet store, are used to pump air through the tea by creating small bubbles. When stones become stained from the tea, clean them between each batch of tea by soaking them in hydrogen peroxide (3% solution that comes from the grocery store). Cleaning the tea maker between every tea batch is also very important.

Worm castings are one of the best composts for making tea, but any quality compost will work. Compost can be put in a nylon tea bag or loose. Larger air stones go on the bottom of the bucket. Small air stones can be put in the tea bags. Five-gallon buckets are good to use and will make about 4 gallons of tea. Add ½ ounce of molasses per gallon of water to help feed the microbes. Too much molasses can destroy microbes in the tea. Look at the movement of the water to make sure you have plenty of air and water movement. Put a lid on the bucket and let it brew for six or eight hours. These same percentages will work at any scale.

Molasses will provide a sweet smell. When the molasses is used up, the aroma of the tea will change to a yeasty smell. Remove the tea bags if used, and continue to brew the tea with the air pump running for another sixteen to twenty hours. The tea will start to deteriorate immediately after the air pump is turned off. The life of the tea can be extended for a day by leaving the air on, but when all the food has been used up, it will deteriorate even with the air on. Never try to store your finished tea in a closed container. It will develop pressure inside and blow.

Five gallons of tea will cover an acre of planting. As a soil drench, 5 gallons will cover about 10,000 square feet of lawn or garden. It doesn't matter how much water you use to dilute and spread the tea.

Cornmeal Juice

Cornmeal juice is a natural fungal control for use in any kind of sprayer. Make by soaking whole ground cornmeal in water at 1 cup per 5 gallons of water. Put the cornmeal in a nylon stocking or other permeable bag to hold in the larger particles. If used loose, cornmeal solids should be strained out before use. The milky juice of the cornmeal will permeate the water, and this mix should be sprayed without further diluting. Cornmeal juice can be mixed with compost tea, Garrett Juice, or any other natural foliar-feeding spray.

Detox Program for Contaminated Soil

Digging the soil out and hauling it off is not the answer. That just moves the problem from point A to point B.

If your soil has been contaminated with heavy metals like arsenic and chromium in treated lumber or creosote in railroad ties, or with lead and arsenic from iron supplements, or if the contamination is from pesticides or petroleum spills, the solution is the same. First, stop the contamination. Second, apply the activated charcoal product from Norit called Gro-Safe. It's very fine-textured and must be mixed with water to apply. Since it is hard to find, fine-textured humate can be substituted.

The next step is to drench the problem area with the Garrett Juice solution plus orange oil. Use the Garrett Juice formula above and add 2 ounces of orange oil or D-limonene per gallon of mix. Gro-Safe will tie up the contaminants, and the Garrett Juice and orange oil will stimulate the microbes to eat the contaminants. Liquid molasses is in the Garrett Juice mix, but adding dry molasses to the soil at 10–20 pounds per 1,000 square feet will greatly help the decontamination process.

Fire Ant Mound Drench

Mix equal amounts of compost tea, molasses, and orange oil. Use 4–6 ounces of this concentrate per gallon of water and use as a drench to kill fire ants and other pests in the ground. For fire ants, use a container that pours a solid stream of liquid and pour into one spot in the center of the mound. This causes the mix to go to the bottom of the mound for best kill of the queens. Then pour the remainder of the mix in a circular pattern covering the entire mound.

Garlic Tea

To make garlic tea, simply omit the pepper from the Garlic-Pepper Tea recipe below and add another bulb of garlic. In a blender with water, liquefy three bulbs of garlic. Strain away the solids. Pour the garlic juice into a 1-gallon container. Fill the remaining volume with water to make 1 gallon of concentrate. Shake well before using and add ¼ cup of the concentrate to each gallon of water in the sprayer.

For additional power, add one tablespoon each of seaweed and molasses to each gallon or use the entire Garrett Juice mixture. Always use plastic containers with loose-fitting lids for storage if needed.

Garlic-Pepper Tea

An organic insect and disease control material made from the juice of garlic and hot peppers such as jalapeño, habanero, or cayenne. This is basically a preventative control. However, its use should be limited because it will kill small beneficial insects. It is effective for both ornamental and food crops.

In a blender with water, liquefy two large bulbs of garlic and two cayenne or habanero peppers. Strain away the solids. Pour the garlic-pepper juice into a 1-gallon container. Fill the remaining volume with water to make one gallon of concentrate. Shake well before using and add ¼ cup of the concentrate to each gallon of water in the sprayer.

Garrett Juice

Garrett Juice is a foliar-feeding liquid, a root stimulator, a soil amendment, and an organic fertilizer, and it is excellent for setting buds and producing flowers.

Ready-to-use Garrett Juice

Mix the following in 1 gallon of water:

1 cup compost tea or liquid humate
1 ounce molasses

1 ounce apple cider vinegar
1 ounce liquid seaweed

Garrett Juice Concentrate

Mix the following: 1 gallon of compost tea or liquid humate, 1 pint liquid seaweed, 1 pint apple cider vinegar, and 1 pint molasses. Use 1½ cups of concentrate per gallon of water.

Garrett Juice Plus (for more fertilizer value)

Add 1–2 ounces of liquid fish (fish hydrolysate) per gallon of Garrett Juice spray.

For disease and insect control, add ¼ cup garlic tea, ¼ cup garlic-pepper tea, or 1–2 ounces orange oil.

Garrett Juice Plus Concentrate

Add 1 pint fish hydrolysate. Use 1–1½ cups of the concentrate per gallon of water for the spray.

Orange Oil

Orange oil contains the raw oil collected from the citrus peel during the juicing extraction. No heat is applied during this "cold-pressed" process, thereby preserving the integrity of the oil. Orange oil degrades the waxy coating on the exoskeleton of insects, causing dehydration and asphyxiation.

Foliar Solution: Mix 2 ounces per gallon of water. Spray plants during the cooler part of the day to prevent burning of foliage. Works on a wide range of insect pests.

Soil Drench: Mix 2 ounces per gallon of water. Diluted solution may be poured directly on soil as a natural control for a variety of mound-dwelling insects.

Cleaning Solution: Mix 2 ounces per gallon of water. Diluted solution may be used as a natural household cleaner. Test on a small area before use.

Organic Pecan and Fruit Tree Program

Pecan trees and most other fruit trees can be grown organically with great success. Trees should never have bare soil in the root zone, but should always be covered with native shredded mulches and/or native grasses, legumes, or other cover crops. Trees should be planted high with natural organic techniques. Trunk flares should be easily visible. Exposing the flares properly is the most important first step.

Soil Feeding Schedule (first year)

Round 1—February: Mechanically aerate root zone and apply organic fertilizer at 20 pounds per 1,000 square feet, whole ground cornmeal at 20 pounds per 1,000 square feet, dry molasses at 20 pounds per 1,000 square feet, and zeolite at 20 pounds per 1,000 square feet.

Round 2—June: Apply organic fertilizer at 10 pounds per 1,000 square feet and greensand at 40 pounds per 1,000 square feet. Use soft rock phosphate or other rock mineral product at 40 pounds per 1,000 square feet.

Round 3—September: Apply organic fertilizer at 10 pounds per 1,000 square feet and Sul-Po-Mag at 20 pounds per 1,000 square feet. Clean wood charcoal can also be used. See Terra Petra on DirtDoctor.com. To refine the program, send soil samples for analysis and recommendations. Be sure to add a note that your program will be organic and that test results should be sent to the Natural Organic Warehouse for additional recommendations based on the soil tests.

Spray Schedule

1st spraying: Spray Garrett Juice Plus and Bio Wash, drench with Garrett Juice Plus and Tree and Shrub THRIVE at pink bud.
2nd spraying: After flowers have fallen. For best results, spray every two weeks, but at least once a month.
3rd spraying: About June 15th (later in northern locations).
4th spraying: Last week in August through mid-September. Use additional sprayings as time and budget allow.
Note: If other quality organic sprays are used, make sure apple cider vinegar is included at 1 ounce per gallon.

Alternate Program (after first year or for limited-budget projects)

1. Apply dry molasses at 20 pounds per 1,000 square feet.
2. Spray Garrett Juice Plus and Bio Wash every thirty days. Treat root zone with Tree and Shrub THRIVE.

Pruning

Very little pruning is needed or recommended. Maintain cover crops and/or natural mulch under the trees year-round. Never cultivate the soil under pecan and fruit trees.

Insect Release

Trichogramma wasps: Weekly releases of 10,000–20,000 eggs per acre or residential lot starting at bud break and continuing for three weeks.
Green lacewings: Release 4,000 eggs per acre or residential lot weekly for one month.
Ladybugs: Optional and as needed, release 1,500–2,000 adult beetles per 1,000 square feet when shiny honeydew is seen on foliage.

Organic Rose Program

Roses are actually easy to grow and should only be grown organically, since they are one of the best medicinal and culinary herbs in the world. When they are loaded with toxic pesticides and other chemicals, this use is gone, or at least it should be. Drinking rose hip tea or using rose petals in teas or salads sprayed with synthetic poisons is a bad idea. For best results with roses of any kind, here's the organic program that really works.

Selection

Buy and plant well-adapted roses for your area. Old roses will have the largest and most vitamin C–filled hips. They are also the most fragrant and the best-looking bushes for landscape use. However, any roses can be grown with this program.

Planting

Prepare rose beds by mixing the following into existing soil to form a raised bed: 6" compost, ½" lava sand, ½" decomposed granite, 20 pounds dry molasses, 20 pounds whole ground cornmeal, 30 pounds zeolite, and 20 pounds Sul-Po-Mag per 1,000 square feet. Remove all the soil from the root balls and soak the bare roots in water with Garrett Juice and one of the mycorrhizal fungi products such as THRIVE per label directions. Spread roots radiating out from the main stem and settle the soil around plants with water. For root stimulation, Garrett Juice and THRIVE can be added to the soaking water or when watering in the plants. Use the products at the same mixing rates used for spraying.

Mulching

After installing the plants, cover all the soil in the beds with ½–1 inch of compost or earthworm castings, followed by 2–3 inches of shredded native tree trimmings mulch. Do not pile the mulch up on the stems of the roses.

Watering

If possible, save and use rainwater for irrigation. If not, add 1 tablespoon of apple cider vinegar per gallon of water or Garrett Juice at half the spray mix rate.

Feeding Schedule (first year)

Round 1: Apply organic fertilizer or dry molasses at 20 pounds per 1,000 square feet, zeolite at 40 pounds per 1,000 square feet, and horticultural cornmeal at 10–20 pounds per 1,000 square feet.

Round 2: Apply organic fertilizer at 20 pounds per 1,000 square feet, greensand at 40 pounds per 1,000 square feet, soft rock phosphate or Flora Stim at 30 pounds per 1,000 square feet.

Round 3: Apply organic fertilizer such as alfalfa meal at 20 pounds per 1,000 square feet and Sul-Po-Mag at 20 pounds per 1,000 square feet.

Feeding Schedule (after the first year or for limited-budget projects)

1. Apply dry molasses at 20 pounds per 1,000 square feet.
2. Spray Garrett Juice Plus and Bio Wash every thirty days. Treat root zone with Tree and Shrub THRIVE.

Pest Control

Apply dry granulated garlic to the soil.

For insect pest and disease control in general, spray roses every thirty days with Garrett Juice and garlic tea or Bio Wash.

For thrips, apply beneficial nematodes to the soil in early spring or when foliage begins to grow.

Sick Tree Treatment

Trees that are adapted to your site should grow well and not have troubles. Trees become infested with insect pests, parasites, and diseases when they are in stress and sick. Mother Nature then sends in the cleanup crews. Pests are just doing their jobs—trying to take out the sick, unfit plants. Most plant sickness is environmental: too much water; not enough water; too much fertilizer; wrong kind of fertilizer; toxic chemical pesticides; soil compaction; grade changes; ill-adapted plant variety selection; and/or planting too many of the same plants, creating monocultures, as was done with American elms in the Pacific Northwest and the red oak/live oak communities in certain parts of the South.

The natural organic tree health plan is simple. Keep trees in a stress-free condition so their immune systems can resist insect pests and diseases. For example, it has been noticed by many farmers and ranchers that oak wilt doesn't bother some trees, especially those that have properly exposed root flares, have not had synthetic fertilizers and herbicides used on them, and whose natural habitat under them has been maintained. This process is not just good for oak wilt but for preventing and curing many other tree problems as well. Here is the updated version and how it works.

Step 1: Stop Using ALL High-Nitrogen Fertilizers and Toxic Chemical Pesticides

Toxic chemical pesticides kill beneficial nematodes, other helpful microbes, and good insects and also control the pest insects poorly. Synthetic fertilizers are unbalanced, harsh salts and are often contaminated and destructive to the chemistry, the physics, and the life in the soil. They feed plants poorly and contaminate the environment. They volatilize into the air, wash away, and leach through the soil into the water system.

Step 2: Remove Excess Soil from the Trunk Flare

A very high percentage of trees are too deep in their containers, have been planted too low, or have had fill soil or eroded soil added on top of the trunk flares. Soil or even heavy mulch covering the trunk flares blocks oxygen, keeps bark moist, and leads to circling and girdling roots. Ideally, excess soil and circling and girdling roots should be removed before planting. Removing soil from the trunk flares of planted trees should be done professionally with a tool called the Air Spade or Air Knife. Homeowners can do the work by hand with a stiff brush, gentle water, and a Shop-Vac or even a power washer if done very carefully. Vines and groundcovers should also be kept off tree trunks and pruned back away from the flares, at least on an annual basis.

Step 3: Aerate the Root Zone Heavily

Don't rip, till, or plow the soil around trees. That destroys all the feeder roots. Punch holes (with turning forks, core aerators, or agriculture devices such as the Air-Way) heavily throughout the root zone. Liquid injectors or Air Spade–type tools can also be used. Start between the dripline and the trunk and go far out beyond the dripline. For a normal-sized residential property, the entire site should be done. Holes 6–8 inches deep are ideal, but any depth is beneficial.

Step 4: Apply Organic Amendments

Apply zeolite 40–80 pounds per 1,000 square feet, greensand at 40–80 pounds per 1,000 square feet, lava sand at 80–120 pounds per 1,000 square feet, whole ground cornmeal at 20–30 pounds per 1,000 square feet, and dry molasses at 10–20 pounds per 1,000 square feet. Cornmeal is a natural disease fighter, and molasses is a carbohydrate source to feed the microbes in the soil. The rock materials provide structural improvement and minerals. Finish with a 1-inch layer of compost, followed by a 3-inch layer of shredded native tree trimmings in bare areas; however, do not pile mulch up on the root flare or the trunk. Smaller amounts of these materials can be used where budget restrictions exist. The application can also be done at separate times. Also, any rock dust material different from the base rock on the site will help.

Step 5: Spray Trees and Soil

Spray the ground, trunks, limbs, twigs, and foliage of trees with the entire Garrett Juice mixture. Do this monthly or more often if possible. For large-scale farms and ranches, a one-time spraying is beneficial if the budget doesn't allow ongoing sprays. Adding garlic oil, cornmeal juice, or Bio Wash to the spray is also beneficial for disease control while the tree is in trouble. Cornmeal juice is a natural fungal control that is made by soaking horticultural or whole ground cornmeal in water at 1 cup per 5 gallons of water (see formula above). Screen out the solids and spray without further dilution. Cornmeal juice can be mixed with Garrett Juice or any other natural foliar-feeding spray. It can also be used as a soil drench for the control of soil-borne diseases. Dry granulated garlic can also be used on the soil in the root zone at about 1–2 pounds per 1,000 square feet for additional disease control. Adding Bio Wash to the spray will help protect against insect pests and disease pathogens and encourage plant growth.

During drought conditions, adding moisture to the soil is a critical component of the Sick Tree Treatment.

Tree Trunk Goop

Mix equal thirds of each in water: diatomaceous earth, soft rock phosphate, compost. Paint or slather onto cuts, borer holes, or other injuries on trunks or limbs. Reapply if washed off by rain or irrigation. In the soil, it makes a good organic fertilizer.

Vinegar

Vinegar Fungicide

Mix 3 tablespoons of natural apple cider vinegar in 1 gallon of water. Spray during the cool part of the day for black spot on roses and other fungal diseases. Adding molasses at 1 tablespoon per gallon will also help.

Vinegar Herbicide

The best choice for herbicide use is 10 percent white vinegar made from grain alcohol. It should be used full strength. I've mentioned 20 percent in the past, but it is stronger than needed and too expensive. Avoid products that are made from 99 percent glacial acetic acid. This material is a petroleum derivative. Natural vinegars such as those made from fermenting apples have little herbicidal value. They are used in irrigation water and as an ingredient in Garrett Juice.

Herbicide Formula

1 gallon of 10 or 20 percent vinegar
1 ounce orange oil or D-limonene
1 tablespoon molasses
1 teaspoon liquid soap or other surfactant (I use Bio Wash)
Do not add water

Shake well before each spraying and spot-spray weeds. Keep the spray off desirable plants. This spray will injure any plants it touches. This natural spray works best on warm to hot days.

Vinegar sprayed on the bases of trees and other woody plants will not hurt the plant at all. This technique was first learned about by spraying the suckers and weeds growing around the bases of grapevines. For best results, use full-strength 10 percent (100 grain) vinegar with 1 ounce orange oil and 1 teaspoon liquid soap per gallon. Avoid vinegar products made from acetic acid.

Appendix 2. Sources for Organic Supplies

Biological Inoculant Sources

Alpha BioSystems, Inc. (microbe products)
9912 W. York Street, Wichita, KS 67215
(888) 265-7929 or (316) 265-7929
http://www.alphabiosystems.com/main.cfm?menu=default

BioOrganics (commercial only)
Mycorrhizal Inoculants
P.O. Box 5326, Palm Springs, CA 92263
(888) 332-7676
http://www.bio-organics.com

Bio-S.I. Technology, LLC.
P.O. Box 784, Argyle, TX 76226
(866) 393-4786
http://www.biositechnology.com/

Mycorrhizal Applications
P.O. Box 1029, Grants Pass, OR 97528
(541) 476-3985; (866) 476-7800
http://www.mycorrhizae.com

T&J Enterprises
2328 W. Providence Avenue, Spokane, WA 99205
(509) 327-7670
http://www.tandjenterprises.com

Mail-Order Sources

ARBICO (Beneficial insects, organic supplies, baits, etc.)
Arizona Biological Controls
P.O. Box 8910, Tucson, AZ 85738
(800) 827-BUGS (2847)
http://www.arbico-organics.com/

BioLogic Company (Beneficial nematodes)
P.O. Box 177, Willow Hill, PA 17271
(717) 349-2789
http://www.biologic.com

BioPac Crop Care, Inc. (Home garden insect control)
P.O. Box 87, Mathis, TX 78368
(361) 547-3259
http://www.biofac.com

Gardener's Supply Company
128 Intervale Road, Burlington, VT 05401-2804
(888) 833-1412
http://www.gardeners.com

Green Living (Natural organic products for the home or garden)
(214) 821-8444
http://www.green-living.com

Green Methods Catalog (Supplier of beneficial insects and biological control products; also in small quantities for use in home gardens)
The Green Spot, Ltd.
93 Priest Rd., Nottingham, NH 03290-6240
(603) 912-8925
http://www.greenmethods.com

Harmony Farm Supply and Nursery (Insects and organic supplies)
P.O. Box 460, Graton, CA 95444
(707) 823-9125
http://www.harmonyfarm.com

Integrated BioControl Systems (Beneficial nematodes)
P.O. Box 96, Aurora, IN 47001
(812) 537-8674
http://www.goodbug-shop.com

KUNAFIN (Insects)
Rt. 1, Box 39, Quemado, TX 78877
(800) 832-1113
http://www.kunafin.com

M&R Durango (Beneficial nematodes)
P.O. Box 886, Bayfield, CO 81122
(800) 526-4075
http://www.goodbug.com

Natural Insect Control (Insects and organic supplies)
RR #2, Stevensville, ON, LOS 1SO Canada
(905) 382-2904
http://www.naturalinsectcontrol.com

Natural Organic Warehouse
P.O. Box 676, Andover, KS 67002
(888) 998-9969
www.naturalorganicwarehouse.com

Nature's Control (Insects and some supplies)
P.O. Box 35, Medford, OR 96501
(541) 245-6033
http://www.hiredbugs.com

Peaceful Valley Farm Supply (Insects and organic supplies)
P.O. Box 2209, Grass Valley, CA 95945
(916) 272-4769
www.groworganic.com

Planet Natural (Insects and supplies)
1612 Gold Avenue, Bozeman, MT 59715
(800) 289-6656
www.planetnatural.com

Rincon-Vitova Insectaries (Insects)
P.O. Box 1555, Ventura, CA 93002
(805) 643-5407
www.rinconvitova.com

Wholesale Distributors

National Distribution:

Natural Organic Warehouse (Wholesale distribution of organic products)
P.O. Box 676, Andover, KS 67002
(888) 998-9669
http://www.naturalorganicwarehouse.com

Alabama:		
BWI	Jackson, MS	601-922-5214
GroSouth	Montgomery	334-265-8241
Arizona:		
Excel		800-479-0121
Arkansas:		
BWI-Memphis	Memphis, TN	901-367-2941
BWI-Springfield	Springfield, MO	417-881-3003
BWI-Texarkana	Nash, TX	903-838-8561
Hummert		800-325-3055
California:		
Excel CA		800-479-0121
Excel Garden Products	Santa Fe	562-952-1234
Excel Garden Products	Sacramento	800-535-3100
Colorado:		
Excel Portland		800-422-7008
Western Plains		800-700-8722
Connecticut:		
Arett		800-257-8220
Commerce		800-289-0982

Delaware		
Arett		800-257-8220
Commerce		800-289-0982
Florida		
BWI		800-289-0982
JR Johnson		800-553-8310
Georgia		
BWI		800-289-0982
GroSouth	Tucker	800-282-3682
Idaho		
Excel Portland		800-422-7008
Illinois		
Commerce		800-289-0982
Hummert		800-325-3055
Olsen		847-381-9333
Prince		800-777-2486
Siemer		800-747-0074
VFC Distributors, Inc.	Milan	309-787-1749
Indiana		
Commerce		800-289-0982
Prince		800-777-2486
Siemer		800-747-0074
Waldo		800-468-4011
Iowa		
Hummert		800-325-3055
Northwest Feeds		402-571-0305
JR Johnson		800-328-9221
Kansas		
Hummert		800-325-3055
BWI-Kansas City	Lenexa	913-859-9009
Garden Wise	Wichita	800-362-3033
Kentucky		
Commerce		800-289-0982

Louisiana		
BWI-Jackson	Jackson, MS	601-922-5214
BWI-Texarkana	Nash, TX	903-838-8561
Maine		
Arett		800-257-8220
Commerce		800-289-0982
Maryland		
Arett		800-257-8220
Commerce		800-289-0982
Massachusetts		
Arett		800-257-8220
Commerce		800-289-0982
Michigan		
Commerce		800-289-0982
Prince		800-777-2486
Waldo		800-468-4011
Minnesota		
Prince		800-777-2486
JR Johnson		800-328-9221
Mississippi		
BWI-Jackson		601-922-5214
BWI-Memphis	Memphis, TN	901-367-2941
Missouri		
BWI-Springfield		417-881-3003
Hummert		800-325-3055
Siemer		800-747-0074
Montana		
Excel Portland		800-422-7008
Nebraska		
BWI-Kansas City	Lenexa, KS	800-859-9009
JR Johnson		800-328-9221

Nevada	
Excel CA	800-479-0121
New Hampshire	
Arett	800-257-8220
Commerce	800-289-0982
New Jersey	
Arett	800-257-8220
Commerce	800-289-0982
New Mexico	
Greenhouse & Garden Supply	505-345-8989
Western Plains	800-700-8722
New York	
Arett	800-257-8220
Commerce	800-289-0982
North Carolina	
Arett	800-257-8220
Commerce	800-289-0982
North Dakota	
JR Johnson	800-328-9221
Ohio	
Arett	800-257-8220
Commerce	800-289-0982
Prince	800-777-2486
Waldo	800-438-4011
Oklahoma	
Ecoh Distributing	800-634-5296
BWI-Springfield	417-881-3003
Oregon	
Excel Portland	800-422-7008
Gard'N-Wise Dist.	503-655-1475
Excel Garden Prod.	503-650-4000

Pennsylvania

Arett	800-257-8220
Commerce	800-289-0982

Rhode Island

Arett	800-257-8220
Commerce	800-289-0982

South Carolina

BWI-Greenville/Spartanburg	800-922-8961

South Dakota

JR Johnson	800-328-9221

Tennessee

BWI-Memphis	901-267-2941

Texas

BWI-Dallas	913-242-4755
Excel Garden Products	972-660-2999
Nitro-Phos Fertilizers	713-228-1868
Natural Organic Warehouse	888-998-9669
San Jacinto Environmental Supplies	713-957-0909
Gard'N-Wise	806-744-8894
Adams Wholesale Supply	210-822-3141
Coastal Ag Supply	281-943-1450
BWI-Texarkana	903-838-8561

Utah

Excel Portland	800-422-7008
Western Plains	800-700-8722

Vermont

Arett	800-257-8220
Commerce	800-289-0982

Virginia

Arett	800-257-8220
Commerce	800-289-0982

Washington
Excel Portland 800-422-7008

West Virginia
Arett 800-257-8220
Commerce 800-289-0982

Wisconsin
Commerce 800-289-0982

Wyoming
Prince 800-777-2486
JR Johnson 800-328-9221
Excel Portland 800-422-7008
Western Plains 800-700-8722

Appendix 3. Soil-Testing Resources

Soil-Testing Laboratories (Biological)

BBC Laboratories
1217 North Stadem Drive, Tempe AZ 85281
(480) 967-5931 FAX (480) 967-5036
www.bbc-labs.com

Microbial Matrix Systems, Inc.
33935 Hwy 99E, Suite B, Tangent, OR 97389
(541) 967-0554 FAX (541) 967-4025

Soil Foodweb, Inc.
1128 NE 2nd street, Suite 120, Corvallis, OR 97330
(541) 752-5066 Fax (541) 752-5142 (Elaine Ingham, PhD)
www.soilfoodweb.com

Soil Foodweb New York, Inc. (Biological and Chemical)
555 Hallock Ave. (Rte. 25a), Suite 7, Port Jefferson Station, NY 11776
(631) 474-8848 FAX (631) 474-8847

Soil-Testing Laboratories (Chemical)

Peaceful Valley Farm Supply
P.O. Box 2209 Grass Valley, CA 95945
(916) 272-4769
www.tpsl.biz

Soil and Plant Laboratory
352 Mathew Street, Santa Clara, CA 95050
(408) 727-5125 (James West)

Soil Control Lab
42 Hanger Way, Watsonville, CA 95076
(408) 724-5422 (Frank Shields)

Texas Plant and Soil Lab (specializes in organic production techniques)
5115 W. Monte Cristo, Edinburg, TX 78539
(956) 383-0739

Tiberleaf Soil Testing Services
26489 Ynez Rd., Suite C-197, Temecula, CA 92591
(909) 677-7510

Wallace Laboratories
365 Coral Circle, El Segundo, CA 90245
(310) 615-0116

Appendix 4. Conversion Tables

Fertilizer Conversions

Rate per Acre (pounds)	*Pounds per 1,000 Square Feet*	*Pounds per Square Foot*
100	2½	¼
200	5	½
400	9	1

Manure and Compost Application Rates

Rate per Acre (pounds)	*Pounds per 1,000 Square Feet*	*Pounds per Square Foot*
4 Tons	200	20
8 Tons	400	40

One Cubic Yard Equals

1,296 square feet	¼" deep
648 square feet	½" deep
324 square feet	1" deep
108 square feet	3" deep
81 square feet	4" deep
54 square feet	6" deep
40½ square feet	8" deep
27 square feet	12" deep

Application Rate Chart

800 lb./acre	=	20 lbs./1,000 sq. ft.
400 lb./acre	=	10 lbs./1,000 sq. ft.
250 lb./acre	=	6 lbs./1,000 sq. ft.
200 lb./acre	=	5 lbs./1,000 sq. ft.
1 qt./acre	=	2 tbsp./1,000 sq. ft. = 1 oz./1000 sq. ft.
13 oz./acre	=	1 tsp./1000 sq. ft. = 3 oz./1000 sq. ft.
11 gal./acre	=	1 qt./1,000 sq. ft.
1 qt./acre	=	2 tbsp./1,000 sq. ft.
1 lb./acre	=	.4 oz./1,000 sq. ft.
1.5 oz./acre	=	7 drops/gal./1000 sq. ft.
13 oz./acre	=	1 tsp./1,000 sq. ft.
4 oz./acre	=	.10 oz./1,000 sq. ft. (30 drops per gal.)
8 oz./acre	=	.20 oz./1,000 sq. ft. (60 drops per gal.)
1 gal./10 acres	=	13 oz./acre
1.5 gal./10 acres	=	19 oz./acre
.5 gal./10 acres	=	6 oz./acre
6 in. soil	=	2 million lbs./acre

General Measurement Tables

Common Measurements

One pinch or dash	=	1/16 teaspoon
1 teaspoon	=	1/6 ounce or 60 drops
1 tablespoon	=	180 drops
1 ounce	=	360 drops
1 tablespoon	=	3 teaspoons (1 ounce liquid, 180 drops)
1 gallon	=	769 teaspoons
4 tablespoons	=	¼ cup (2 ounces liquid)
⅓ cup	=	5 tablespoons plus 1 teaspoon
½ cup	=	8 tablespoons (4 ounces liquid)
1 gill	=	½ cup (4 ounces liquid)
1 cup	=	16 tablespoons (8 ounces liquid)
1 pint	=	2 cups (16 ounces liquid)
1 quart	=	2 pints (32 ounces liquid)
1 gallon	=	4 quarts
1 gallon	=	16 pints
1 gallon	=	32 cups
1 peck	=	8 quarts
1 bushel	=	4 pecks
1 pound	=	16 ounces (dry measure)

Linear Measure

1 foot	=	12 inches
1 hand	=	⅓ foot = 4 inches
1 span	=	9 inches
1 yard	=	3 feet
1 rod	=	16½ feet = 5½ yards
1 furlong	=	40 poles = 220 yards
1 mile	=	8 furlongs = 1,760 yards = 5,280 feet = 320 rods
1 league	=	3 miles
1 degree	=	69⅛ miles

Square or Area Measure

1 square foot	=	144 square inches
1 square yard	=	9 square feet
1 acre	=	160 square rods = 43,560 sq. ft.
1 section	=	640 acres = 1 square mile
1 hectare	=	2.47 acres

Common Metric-U.S. Equivalents

Unit	*Metric Equivalent*	*U.S. Equivalent*
Acre	0.405 hectares	43,560 sq. ft.
Barrel (petroleum, U.S.)	158.99 liters	42 gallons
Bushel	35.24 liters	4 pecks
Cord (wood)	3.63 cu. meters	128 cu. ft.
Cubic Foot	28.32 liters	7.48 gallons
Cubic Inch	16.387 cu. centimeters	.000578 cu. ft.
Cubic Meter	1,000 liters	1.308 cu. yd.
Cubic Yard	.765 cu. meters	27 cu. ft.
	.765 liters	201.974 gallons
Cup	0.24 liters	8 ounces, liquid
Degrees, Celsius	Water boils at 100°C, freezes at 0°C	Multiply by 1.8 and add 32 to obtain F
Degrees, Fahrenheit	Subtract 32 and divide by 1.8 to obtain C	Water boils at 212°F, freezes at 32°F
Foot	30.48 centimeters	12 inches
Furlong	201.17 meters	220 yards
Gallon, liquid (U.S.)	3.79 liters	4 quarts, liquid
Gram	1,000 milligrams	.035 ounces
Hand	10.16 centimeters	4 inches
Hectare	10,000 sq. meters	2.471 acres

Unit	*Metric Equivalent*	*U.S. Equivalent*
Inch	2.54 centimeters	.083 feet
Kilogram	.001 tons, metric	2.204 pounds
Kilometer	1,000 meters	.621 miles
Liter	.001 cu. meters	1,000 milliliters
	61.024 cu. inches	1.057 quarts, liquid
Meter	100 centimeters	1.094 yards
Square mile	258.998 hectares	640 acres

Cubic or Volume Measure

A legal cord of wood is 4 feet high, 4 feet wide, and 8 feet long.

1 cubic foot	=	1,728 cubic inches
1 cubic yard	=	27 cubic feet
1 cord of wood	=	128 cubic feet
1 board foot	=	144 cubic inches = $^{1}/_{12}$ cubic foot

Liquid or Fluid Measure

1 pint (pt.)	=	2 cups
1 quart (qt.)	=	2 pints
1 gallon (gal.)	=	4 quarts
1 barrel (bbl.)	=	31½ gallons
1 acre foot	=	325,000 gallons

Dry Measure

1 quart	=	2 pints
1 peck	=	8 quarts
1 bushel (bu.)	=	4 pecks

Metric Equivalents, Linear

1 millimeter (mm)	=	.0394 in.
1 centimeter (cm)	=	.3937 in.
1 decimeter (dm)	=	3.937 in.
1 meter (m)	=	39.37 in. = 1.1 yard
1 decameter	=	393.7 in = 10 yd. 2.8 ft.
1 hectometer	=	328 ft. 1 in.
1 kilometer	=	3,280 ft. 1 in.

Common Equivalents

1 bushel	=	2,150 cubic inches or 1¼ cubic feet
1 gallon	=	231 cubic inches
1 cubic foot	=	7½ gallons
1 cubic foot of water	=	62½ pounds (62.43 lb.)
1 gallon of water	=	8⅓ pounds (8.345 lb.)
1 cubic foot of ice	=	57½ pounds

Conversions

1 sq. yd.	=	9 sq. ft.
1 cu. yd.	=	27 cu. ft.

Application Rates

Bulk material

(1 cu. ft. =)	(1 cu. yd. =)
12 sq. ft. 1" deep	1,296 sq. ft. ¼" deep
6 sq. ft. 2" deep	648 sq. ft. ½" deep
4 sq. ft. 3" deep	324 sq. ft. 1" deep
3 sq. ft. 4" deep	162 sq. ft. 2" deep
	108 sq. ft. 3" deep
	81 sq. ft. 4" deep

Bagged Material (2 cu. ft. bag)

36 sq. ft. 1" deep	96 sq. ft. ¼" deep
18 sq. ft. 2" deep	48 sq. ft. ½" deep
12 sq. ft. 3" deep	24 sq. ft. 1" deep
9 sq. ft. 4" deep	12 sq. ft. 2" deep
	8 sq. ft. 3" deep
	6 sq. ft. 4" deep

Conversion Tables

U.S.	*Abbreviation*	*Metric*
1 teaspoon = 60 drops	Teaspoon = tsp.	1 teaspoon = 5 milliliters
1 tablespoon = 3 teaspoons	Tablespoon = tbsp.	1 tablespoon = 15 milliliters
1 tablespoon = 180 drops	Cup = c.	1 ounce = 30 milliliters
1 ounce = 2 tablespoons	Pint = pt.	1 quart = .940 liters
1 ounce = 360 drops	Quart = qt.	1 gallon = 3.76 liters
1 cup = 8 ounces	Gallon = gal.	

U.S.	*Abbreviation*
1 pound = 16 ounces	Ounce = oz.
1 pint (16 oz.) = 2 cups	Pound = lb.
1 quart (32 oz.) = 2 pints	Milliliter = mL
1 gallon (128 oz.) = 4 quarts	Liter = L
1 gallon = 16 cups	
1 gallon = 128 ounces	

Dilution Chart

Gallons of Water

Dilution	*1 QT.*	*1 GAL.*	*3 GAL.*	*5 GAL.*	*10 GAL.*	*15 GAL.*
1-10	3 oz.	12 oz.	2¼ pt.	2 pts.	3¾ qt.	5½ qt.
1-50	4 tsp.	5 tbsp.	7½ oz.	12½ oz.	25 oz.	37½ oz.
1-80	1 tbsp.	2 oz.	6 oz.	10 oz.	20 oz.	20 oz.
1-100	2 tsp.	2½ tbsp.	3½ oz.	6¼ oz.	12½ oz.	19 oz.
1-200	1 tsp.	4 tsp.	2 oz.	3½ oz.	6½ oz.	10 oz.
1-400	½ tsp.	2 tsp.	2 tbsp.	1½ oz.	3 oz.	5 oz.
1-800	—	1 tsp.	1 tbsp.	5 tsp.	1½ oz.	2½ oz.

Soil Nutrient Availability

	Low	*Normal*	*High*	*Very High*
Calcium	<20	20–60	60–80	>80
Magnesium	<10	10–25	25–35	>35
Potassium	<5	5–20	20–30	>30
Phosphorus	<.1	.1–.4	.5–.8	>.8
Nitrogen	<1	1–10	10–20	>20
Nitrate	<5	5–50	50–100	>100
Sulfate	<30	30–90	90–180	>180
Sulfur	<10	10–30	30–60	>60

Note: Numbers above represent available percentages in the soil.

Convenient Conversion Factors

Multiply	*By*	*To Get*
Acres	43,560	Square feet
Acres	160	Square rods
Acres	4840	Square yards
Bushels	2150.42	Cubic inches
Bushels	2	Pecks
Bushels	64	Pints

Multiply	*By*	*To Get*
Bushels	32	Quarts
Centimeters	0.3937	Inches
Centimeters	0.01	Meters
Centimeters	10	Millimeters
Cubic Centimeters	0.03382	Ounces (liquid)
Cubic Feet	1,728	Cubic Inches
Cubic Feet	0.03704	Cubic Yards
Cubic Feet	7.4805	Gallons
Cubic Feet	59.84	Pints (liquid)
Cubic Feet	29.92	Quarts (liquid)
Cubic Inches	0.000465	Bushels
Cubic Inches	16.39	Cubic Centimeters
Cubic Inches	0.004329	Gallons
Cubic Inches	0.5541	Ounces (liquid)
Cubic Inches	0.02976	Pints (dry)
Cubic Inches	0.0346	Pints (liquid)
Cubic Inches	0.01488	Quarts (dry)
Cubic Inches	0.0173	Quarts (liquid)
Cubic Meters	1,000,000	Cubic Centimeters
Cubic Meters	35.31	Cubic Feet
Cubic Meters	61,023	Cubic inches
Cubic Meters	1.308	Cubic yards
Cubic Meters	264.2	Gallons
Cubic Meters	2,113	Pints (liquid)
Cubic Yards	27	Cubic Feet
Cubic Yards	46,656	Cubic Inches
Cubic Yards	0.7646	Cubic Meters
Cubic Yards	202	Gallons
Cubic Yards	1,616	Pints (liquid)
Cubic Yards	807.9	Quarts (liquid)
Feet	30.48	Centimeters
Feet	12	Inches
Feet	0.3048	Meters
Feet	0.060606	Rods
Feet	⅓ or 0.3333	Yards
Feet per minute	0.01667	Feet per second
Feet per minute	0.01136	Miles per hour
Gallons	3,785	Centimeters
Gallons	0.1337	Cubic Feet
Gallons	231	Cubic Inches
Gallons	128	Ounces (liquid)
Gallons	8	Pints (liquid)
Gallons	4	Quarts (liquid)

Multiply	*By*	*To Get*
Gallons of water	8.3453	Pounds of water
Grains	0.0648	Grams
Grams	15.43	Grains
Grams	0.001	Kilograms
Grams	1,000	Milligrams
Grams	0.0353	Ounces
Grams per liter	1,000	Parts per million
Inches	2.54	Centimeters
Inches	0.08333	Feet
Inches	0.02778	Yards
Kilograms	1,000	Grams
Kilograms	2.205	Pounds
Kilometers	3,281	Feet
Kilometers	1,000	Meters
Kilometers	0.6214	Miles
Kilometers	1,094	Yards
Liters	1,000	Cubic Centimeters
Liters	0.0353	Cubic Feet
Liters	61.02	Cubic Inches
Liters	0.001	Cubic Meters
Liters	0.2642	Gallons
Liters	2.113	Pints (liquid)
Liters	1.057	Quarts (liquid)
Meters	100	Centimeters
Meters	3.281	Feet
Meters	39.37	Inches
Meters	0.001	Kilometers
Meters	1000	Millimeters
Meters	1.094	Yards
Miles	5,280	Feet
Miles	63,360	Inches
Miles	320	Rods
Miles	1,760	Yards
Miles per hour	88	Feet per minute
Miles per hour	1.467	Feet per second
Miles per minute	88	Feet per minute
Miles per minute	60	Miles per hour
Ounces (dry)	437.5	Grains
Ounces (dry)	28.3495	Grams
Ounces (dry)	0.0625	Pounds
Ounces (liquid)	1.805	Cubic Inches
Ounces (liquid)	0.0078125	Gallons
Ounces (liquid)	29.573	Milliliters (cubic centimeters)

Multiply	*By*	*To Get*
Parts per million	0.0584	Grains per U.S. gallon
Parts per million	0.001	Grams per liter
Parts per million	8.345	Pounds per million gallons
Pecks	0.25	Bushels
Pecks	537.605	Cubic Inches
Pecks	16	Pints (dry)
Pecks	8	Quarts (dry)
Pints (dry)	0.015625	Bushels
Pints (dry)	33.6003	Cubic Inches
Pints (dry)	0.0625	Pecks
Pints (dry)	0.5	Quarts (dry)
Pints (liquid)	28.875	Cubic Inches
Pints (liquid)	0.125	Gallons
Pints (liquid)	0.4732	Liters
Pints (liquid)	16	Ounces (liquid)
Pints (liquid)	0.5	Quarts (liquid)
Pounds	7,000	Grains
Pounds	453.5924	Grams
Pounds	16	Ounces
Pounds	0.0005	Tons
Pounds of water	0.01602	Cubic Feet
Pounds of water	27.68	Cubic Inches
Pounds of water	0.1198	Gallons
Quarts (dry)	0.03125	Bushels
Quarts (dry)	67.20	Cubic Inches
Quarts (dry)	0.125	Pecks
Quarts (dry)	2	Pints (dry)
Quarts (liquid)	57.75	Cubic Inches
Quarts (liquid)	0.25	Gallons
Quarts (liquid)	0.9463	Liters
Quarts (liquid)	32	Ounces (liquid)
Quarts (liquid)	2	Pints (liquid)
Rods	16.5	Feet
Rods	198	Inches
Rods	5.5	Yards
Square Feet	144	Square Inches
Square Feet	0.11111	Square Yards
Square Inches	0.00694	Square Feet
Square Miles	640	Acres
Square Miles	28,878,400	Square Feet
Square Miles	102,400	Square Rods
Square Miles	3,097,600	Square Yards
Square Rods	0.00625	Acres

Multiply	*By*	*To Get*
Square Rods	272.25	Square Feet
Square Rods	30.25	Square Yards
Square Yards	0.0002066	Acres
Square Yards	9	Square Feet
Square Yards	1,296	Square Inches
Temperature (°C)+17.98	1.8	Temperature °F
Temperature (°F)-32	5/9 or 0.5555	Temperature °C
Ton	907.1849	Kilograms
Ton	32,000	Ounces
Ton	2,000	Pounds
Yards	3	Feet
Yards	36	Inches
Yards	0.9144	Meters
Yards	0.000568	Miles
Yards	0.01818	Rods

Conversion Chart (Metric to English)

Metric Unit	*Multiply by*	*To obtain English unit*
Length		
Millimeters(mm)	0.04	Inches (in.)
Centimeters(cm)	0.4	Inches (in.)
Meters(m)	3.3	Feet (ft.)
Meters(m)	1.1	Yards (yd.)
Kilometers(km)	0.6	Miles (mi.)
Area		
Square Centimeters (cm^2)	0.16	Square inches ($in.^2$)
Square Meters (m^2)	1.2	Square Yards ($yd.^2$)
Square Meters (m^2)	10.8	Square Feet ($ft.^2$)
Square Kilometers (km^2)	0.4	Square Miles ($mi.^2$)
Hectares (ha)	2.5	Acres (ac.)
Mass (weight)		
Grams (g)	0.035	Ounces (oz.)
Kilograms (kg)	2.2	Pounds (lb.)
Metric Tonnes (t)	1.1	Short Tons
Volume		
Milliliters (mL)	0.03	Fluid Ounces (fl. oz.)
Milliliters (mL)	0.06	Cubic Inches ($in.^3$)
Liters (L)	2.1	Pints (pt.)

Metric Unit	*Multiply by*	*To obtain English unit*
Liters (L)	1.06	Quarts (qt.)
Liters (L)	0.26	Gallons (gal.)
Cubic Meters (m^3)	35	Cubic Feet ($ft.^3$)
Cubic Meters (m^3)	1.3	Cubic Yards ($yd.^3$)
Temperature (exact)		
Degrees Celsius (°C)	Multiply by $^9/_5$ then add 32	Degrees Fahrenheit (°F)

Information Resources

Arboretums and Nature Centers

Armand Bayou Nature Center: (281) 474-2551.
Bayou Bend Gardens, Houston, Texas; Bart Brechter, Curator: 713-639-7750.
Edith Moore Nature Sanctuary (Home of Houston Audubon Society): (713) 932-1392.
Houston Arboretum and Nature Center: (713) 681-8433.
Jesse Jones Nature Center: (281) 446-8588.
Mercer Arboretum: (281) 443-8731.
Nature Discovery Center: (713) 667-6550.
Sims Bayou Nature Center: (713) 640-2407.
Stephen F. Austin Mast Arboretum, Nacogdoches, Texas: (281) 342-3034.

Organizations

American Botanical Council (ABC)
P.O. Box 144345, Austin, TX 78714
512-926-4900
www.herbalgram.org

American Environmental Health Center (AEHC)
8345 Walnut Hill, #225, Dallas, TX 75231
214-373-5131

Botanical Research Institute of Texas (BRIT)
509 Pecan St., Ft. Worth, TX 76102
817-332-4441

Center for Holistic Resource Management
P.O. Box 7128, Albuquerque, NM 87194
505-344-3445

Holistic Resource Management (HRM)
1010 Tijeras NW, Albuquerque, NM 87102
505-842-5252

Texas Organic Farmers and Gardening Association (TOFGA)
P.O. Box 1246, Whitney, TX 76602
877-326-5175

Texas Organic Research Center (TORC)
P.O. Box 140650, Dallas, TX 75214
214-365-0606

Publications

Beneficial Insects and Biological Control

Bat Conservation International. Free monthly e-newsletter on bats. (800) 538-2287; http://www.batcon.org.

Biological Control: A Guide to Natural Enemies in North America, by Anthony Shelton. Cornell University, http://www.nysaes.cornell.edu/ent/biocontrol.

Biological Control Virtual Information Center, Department of Entomology, North Carolina State University, http://cipm.ncsu.edu/ent/biocontrol/.

Dead Snails Leave No Trails: Natural Pest Control for the Home Garden, by Loren Nancarrow and Janet Hogan Taylor. Berkeley, CA: Ten Speed Press, 1996.

Department of Entomology, Texas A&M University, http://entowww.tamu.edu/entoweb/.

Horticultural Integrated Pest Management, Texas A&M University, http://hortipm.tamu.edu/pestprofiles/beneficials.html

Insect Management: Biological Control, Department of Entomology, North Carolina State University, http://www.cals.ncsu.edu/entomology/.

Natural Enemies of Vegetable Crop Insect Pests, Virginia Tech University, http://www.hort.vt.edu/faculty/caldwelj/beneficials.html.

Classics

An Agricultural Testament (Oxford University Press, 1940; Rodale Press, 1976) and *The Soil and Health: A Study of Organic Agriculture* (University Press of Kentucky, 2007), both by Sir Albert Howard, are state-of-the-art guides to organics and the use of compost to bring soil back to health. They were written in the 1940s, but are still two of the best publications on the market.

The Albrecht Papers, by William A. Albrecht. Austin, TX: Acres U.S.A., 1992. A compilation of papers by the late Dr. Albrecht that is considered the bible for managing soil health.

Bread from Stones, by Julius Hensel. Austin, TX: Acres U.S.A., 1991. A classic explaining the role of earth minerals in the production of wholesome food crops.

Eco-Farm, An Acres U.S.A. Primer, by Charles Walters. Austin, TX: Acres U.S.A., 1979. One of the best overall guides on organics.

Compost and Compost Tea

ATTRA: Notes on Compost Tea. www.attar.ncat.org/attar-pub/compost-tea-notes.html.

BioCycle. Professional trade magazine dedicated to biological recycling and composting; nontechnical. Available at most libraries with back issues stored on CD-ROM for easy printing of selected articles.

BioCycle Guide to the Art and Science of Composting, edited by the staff of *BioCycle Journal of Waste Recycling*. Emmaus, PA: JG Press, 1991. A semitechnical overview of composting.

"Compost Fundamentals: Considerations before Choosing a Compost Method." Washington State University. Handout at a conference presentation.

Compost Information Web site: Cornell University, www.cfe.cornell.edu/wmi/.

Compost Science & Utilization. A quarterly journal published by the JG Press, Emmaus, Pennsylvania.

Compost Tea Brewing Manual, by Elaine Ingham. www.soilfoodweb.com.

Compost Tea Group. www.groups.yahoo.com/group/compost_tea.

Compost Tea Industry Association. www.composttea.org.

"Compost Tea: Principles and Prospects for Plant Disease Control." *Compost Science and Utilization Journal* (Autumn 2002): 313–338.

Compost Tea Quality: Light Microscope Manual, by Elaine Ingham. www.soilfoodweb.com.

"Fate of Selected Pesticides Applied to Turfgrass: Effect of Composting Residues," by C. Vandervoot, M. Zabik, B. Branham, and D. Lickfeldt. *Bulletin Environmental Contamination and Toxicology* 58 (2000): 38–45; 2000 WSDA Case 003S-00.

"The Herbicide Contaminated Compost Issue," by D. Pittenger and J. Downer. *Landscape Notes* (University of California Cooperative Extension) 16, no. 1 (June 2002).

International Compost Tea Council. www.intlctc.org.

Let It Rot! The Gardener's Guide to Composting, by Stu Campbell. Charlotte, VT: Garden Way Publishing, 1975.

"The Making of Compost Teas: The Next Generation?," by M. Line and Y. Ramona. *BioCycle* 44, no. 12 (2003): 55–56.

Microbiology of Composting, edited by Heribert Insam, Nuntavun Riddech, and Susanne Klammer. Berlin: Springer-Verlag, 2002. A technical reference on composting.

"Occurrence, Degradation and Fate of Pesticides during Composting," by Buyuksonmez Fatih, Robert Rynk, T. Hess, and E. Bechinski. *Compost Science and Utilization*, Part I, 7, no. 4 (1999–): 66–82; Part II, 8, no. 1 (2000–): 61–81.

"Penn State Researchers Uncover Clopyralid in Compost," by N. Houck and E. Burkhart. *BioCycle* (July 2002): 32.

"Persistent Herbicides in Compost," by D. Bezdicek, M. Fauci, D. Caldwell, R. Finch, and J. Lang. *BioCycle* (July 2001): 25.

"Phytotoxicity of Compost Treated with Lawn Herbicides Containing 2,4–D, Dicamba and MCPP," by G. Bugby and R. Saraceno. Connecticut Agricultural Experiment Station, *Bulletin of Environmental Contamination and Toxicology* 52 (1994): 606–611.

The Practical Handbook of Compost Engineering, by Roger T. Haug. Boca Raton, FL: CRC Press, 1993. A technical reference on composting.

The Rodale Book of Composting: Easy Methods for Every Gardener, edited by Deborah L. Martin and Grace Gershuny. Emmaus, PA: Rodale Press, 1992. A nontechnical overview of composting.

Science and Engineering of Composting: Design, Environmental, Microbiological and Utilization Aspects, edited by Harry A. Hoitink and Harold M. Keener. Columbus: Ohio State University, 1993. A technical reference on composting.

The Science of Composting, by Eliot Epstein. Lancaster, PA: Technomic Publishing, 1997. A technical reference on composting.

The Secret Life of Compost, by Malcolm Beck. Austin, TX: Acres, U.S.A., 1997.

"Testing Compost," by M. Watson. Ohio State University Fact Sheet, ANR-15–03. www.ohioonline.osu.edu/anr-fact/0015.html.

"Understanding How Compost Tea Can Control Disease," by S. Scheueren. *BioCycle* (February 2003): 20–25.

Winning the Organics Game: The Compost Marketer's Handbook, by Rodney W. Tyler. Alexandria, VA: American Society for Horticultural Science Press, 1996. The focus is on how to use and sell compost.

Earthworms and Vermiculture

"The Abundance of Earthworms in Agricultural Land and Their Possible Significance in Agriculture," by K. P. Barley. In *Advances in Agronomy*, Vol. 13, edited by A. G. Norman, 249–268. New York: Academic Press, 1961.

The Earth Moved: On the Remarkable Achievements of Earthworms, by Amy Stewart. Chapel Hill, NC: Algonquin Books, 2004. Nontechnical book on earthworms and nature.

Earthworm Ecology and Biogeography in North America, edited by Paul F. Hendrix. Boca Raton, FL: CRC Press, 1995.

"Vermicomposts Suppress Plant Pests and Disease Attacks," by Clive A. Edwards and Norman Q. Arancon. *BioCycle* (March 2004): 51–54.

Vermiculture Technology: Earthworms, Organic Waste, and Environmental Management, edited by Clive A. Edwards, Norman Q. Arancon, and Rhonda L. Sherman. Boca Raton, FL: CRC Press, 2010.

Worm Digest. A nontechnical quarterly news magazine on worms and vermiculture; Box 544, Eugene, Oregon 97440-0544. www.wormdigest.com

Ecology, Environment, Health, and Related Topics

Acres U.S.A.: The Voice of Eco-Agriculture. A monthly publication that is one of the best on eco-agriculture. They also publish many books on alternative agriculture and have many good articles available for reprint in the "toolbox" section of Web site. P.O. Box 91299, Austin, TX 78709; 1-800-355-5313. www.acresusa.com/magazines/magazine.htm.

BUGS Flyer—The Voice of Ecological Urban Horticulture, Biological Urban Gardening Services, P.O. Box 76, Citrus Heights, CA 95611-0076.

"Calculating Carbon Sequestration by Trees." U.S. Dept. of Energy, 1998. Available at http://www.google.com/search?q=us+dept.+of+energy%2C+carbon+sequestration+rates+for+trees%2C+1998&ie=utf-8&oe=utf-8&aq=t&rls=org.mo.

Chemical Exposure and Human Health: A Reference to 314 Chemicals with a Guide to Symptoms and a Directory of Organizations, by Cynthia Wilson. Jefferson, NC: McFarland and Company, 1993.

"Effect of Pesticide-Treated Grass Clippings Used as a Mulch on Ornamental Plants," by B. Branham and D. Lickfeldt. *HortScience* 32, no. 7 (December 1997): 1216–1219.

Empty Harvest, by Bernard Jensen and Mark Anderson. New York: Penguin Putnam, 1990. A nontechnical study of the link between our food, our immune system (health), and our environment.

Fast Food Nation—The Dark Side of the All-American Meal, by Eric Schlosser. New York: Houghton Mifflin, 2001.

Fateful Harvest, by Duff Wilson. New York: Harper Collins, 2002. A history of how hazardous waste is disposed of in synthetic fertilizers and ends up contaminating the food supply.

The Food Revolution: How Your Diet Can Help Save Your Life and Our World, by John Robbins. Berkeley, CA: Conari Press, 2001.

Is This Your Child's World? How You Can Fix the Schools and Homes That Are Making Your Children Sick, by Doris Rapp. New York: Bantam Books, 1997.

Journal of Pesticide Reform. Quarterly published by Northwest Coalition for Alternatives to Pesticides (NCAP), P.O. Box 1393, Eugene, OR 97440, (503) 344-5044.

Lessons in Nature, by Malcolm Beck. Austin, TX: Acres U.S.A., 2005.

Mother Earth News. A magazine dedicated to environmentally friendly ways of living, including gardening. P.O. Box 56302, Boulder, CO 80322-6302, www.motherearthnews.com/.

National Coalition Against the Misuse of Pesticides (NCAMP), 701 E. Street S.E., Suite 200, Washington, D.C. 20003, (202) 543-5450.

News on Earth. A monthly news flyer published by News On Earth Public Concern Foundation, 101 West 23rd. St., PMB 2245, New York, NY 10011, (212) 741-2365.

Our Stolen Future, by Theo Colborn, Dianne Dumanoski, and John Peterson Myers. New York: Dutton, 1996. Web site with info: www.ourstolenfuture.com. Provides an explanation of how chemicals mimic hormones and hurt our health.

Our Toxic Times. A monthly publication of the Chemical Injury Network (also Medical & Legal Briefs: A Referenced Compendium of Chemical Injury), P.O. Box 301, White Sulphur Springs, Montana 59645-0301, (406) 547-2255.

Pesticides and You. A quarterly publication of Beyond Pesticides, http://www.ienearth.org/news/PesticidesandYou.html.

The Rest of the Story . . . about Agriculture Today, by Harold Willis. Published by author, P.O. Box 692, Wisconsin Dells, Wisconsin 53965; revised 1990.

Seeds of Deception, by Jeffrey M. Smith. Portland, ME: Yes! Books, 2003. Documents the significant health dangers of genetically engineered food and the political corruption that allows them on the market; 888-717-7000, www.seedsofdeception.com.

Technical Report. A newsletter published by Beyond Pesticides, National Coalition Against the Misuse of Pesticides (NCAMP), 701 E. Street S.E., Suite 200, Washington, D.C. 20003, (202) 543-5450.

Gardening

The Dirt Doctor's Membership (members-only area of site). Provides information on organic gardening and living, with excellent advice for Texas. P.O. Box 140650, Dallas, TX 75214. www.dirtdoctor.com/

The Dirt Doctor's Guide to Organic Gardening, by Howard Garrett. Austin: University of Texas Press, 1995.

Ecological Golf Course Management, by Paul Sachs and Richard T. Luff. Hoboken, NJ: John Wiley and Sons, 2002.

Gardening Without Work, by Ruth Stout. New York, NY: The Lyons Press, 1998.

"Growth Analysis of Tomatoes in Black Polyethylene and Hairy Vetch Production Systems," by J. R. Teasdale and A. A. Abdul-Baki. *HortScience* 32, no. 4 (1997): 659–663.

"The Guide To Organic Fertilizers," by Vicki Mattern. *Organic Gardening* (May/June 1996): 55–59.

Handbook of Successful Ecological Lawn Care, by Paul D. Sachs. Newbury, VT: Edaphic Press, 1996. Contains easy-to-understand techniques that allow one to have a beautiful lawn without using dangerous chemicals and at a lower cost than with chemical approaches.

Howard Garrett's Texas Organic Gardening, by Howard Garrett. Lanham, MD: Lone Star Books, 1998.

Know It and Grow It: A Guide to the Identification and Use of Landscape Plants, by

Carl E. Whitcomb. Stillwater, OK: Lacebark Publications, 1985.

Managing Healthy Sports Fields, by Paul D. Sachs. Hoboken, NJ: John Wiley and Sons, 2004. A reference for anyone who wants to prevent problems on turfgrass and save time and money. It is up-to-date with the newest, revised methods from the scientific community designed to protect the safety of our children.

The New Seed-Starter's Handbook, by Nancy Bubel. Emmaus, PA: Rodale Books, 1988.

Organic Gardening. A general trade magazine dedicated to organic techniques. Box 7320, Red Oak, IA 51591-0320. Subscriptions: (800) 666-2206; Business Office: (215) 967-5171; www.organicgardening.com.

Organic Manual, by Howard Garrett. Wyomissing, PA: Tapestry Press, 2002. New and revised ed., 2008.

Organic Vegetable and Edible Landscaping, by Howard Garrett. Lanham, MD: Lone Star Books, 1998.

Plant Propagation Made Easy, by Alan Toogood. Portland, OR: Timber Press, 1994.

Plants for Houston and the Gulf Coast, by Howard Garrett. Austin: University of Texas Press, 2008.

Plants of the Metroplex, by Howard Garrett. Rev. ed. Austin: University of Texas Press, 1998.

Seaweed and Plant Growth, by T. L. Senn. Taylors, S.C.: Faith Printing, 1987.

"Snap Bean Production in Conventional Tillage and in No-Till Hairy Vetch," by A. Abdul-Baki and John Teasdale. *HortScience* 32, no. 7 (December 1997): 1191–1193.

Teaming with Microbes: A Gardener's Guide to the Soil Food Web, by Jeff Lowenfels and Wayne Lewis. Portland, OR: Timber Press, 2006.

Texas Bug Book, by C. Malcolm Beck and Howard Garrett. Austin: University of Texas Press, 1999.

Texas Herb Book, by Howard Garrett. Austin: University of Texas Press, 2001.

Texas Tree Book, by Howard Garrett. Lanham, MD: Lone Star Books, 2002.

Year Round Vegetables, Fruits and Flowers for Metro Houston, by Bob Randall. Houston, TX: Year Round Gardening Press, 1999. A resource guide on how to grow plants organically in Houston and the Gulf Coast and where to get the supplies you may need. Sold at many area garden supply stores. Best resource for Houston.

Soil Biology

Biological Approaches to Sustainable Soil Systems, edited by Norman Uphoff et al. Boca Raton, FL: CRC Press, 2006. Available as an e-book at http://www.crcpress.com/product/isbn/9781574445831.

Eco-Farm, An Acres U.S.A. Primer, by Charles Walters. 3rd rev. ed. Austin, TX: Acres U.S.A., 2003. A semitechnical overview of soils.

The Ecology of Soil Decomposition, by Sina M. Adl. Wallingford, UK: CABI, 2003.

Handbook of Microbial Biofertilizers, edited by M. K. Rai. Binghampton, NY: Haworth Press, 2006.

Handbook of Processes and Modeling in the Soil-Plant System, edited by Rolf Nieder and D. Benbi. Binghamton, NY: Haworth Press, 2003.

The Handbook of Trace Elements, by István Pais and J. Benton Jones, Jr. Boca Raton, FL: St. Lucie Press, 1997.

How Soils Work: A Study into the God-Plane Mutualism of Soils and Crops, by Paul W. Syltie. Fairfax, VA: Xulon Press, 2002. Integrates all the latest advancements and information in soil science into one book; presented in a biblical frame of reference.

Humic, Fulvic and Microbial Balance: Organic Soil Conditioning, by William R. Jackson. Evergreen, CO: Jackson Research Center, 1993. Excellent reference book.

Introduction to Soil Physics, by Daniel Hillel. New York: Academic Press, 1982. A technical introduction to soils and soil properties.

Modern Soil Microbiology, edited by Jan Dirk van Elsas, Janet K. Jansson, and Jack T. Trevors. 2d ed. Boca Raton, FL: CRC Press, 2006.

Paramagnetism: Rediscovering Nature's Secret Force of Growth, by Philip Callahan. Austin, TX: Acres U.S.A., 1995.

"The Role of Arbuscular Mycorrhizal Fungi and Glomalin in Soil Aggregation: Comparing Effects of Five Plant Species," by Matthias C. Rillig, Sara F. Wright, and Valeria T. Eviner. *Plant and Soil* 238 (2002): 325–333.

Soil Biology Primer. A USDA booklet available at: www.soils.usda.gov/sqi/soil_quality/soil_biology/soil_biology_primer.html.

Soil Foodweb, by Elaine Ingham. Excellent free electronic newsletter and Web site on biologic methods in gardening and horticulture (also information on compost teas). www.soilfoodweb.com.

Soil Microbiology: An Exploratory Approach, by Mark Coyne. Clifton Park, NY: Delmar Cengage Learning, 1999.

Soil Microbiology, Ecology, and Biochemistry, edited by Eldor A. Paul. 3rd ed. New York: Academic Press, 2007.

Soil Mineralogy with Environmental Applications, edited by Joe Boris Dixon and Darrell G. Schulze. Soil Science Society of America Book Series, No. 7. Madison, WI: Soil Science Society of America, 2002.

Start with the Soil, by Grace Gershuny. Emmaus, PA: Rodale Press, 1993. A nontechnical overview of soils.

Sustainable Soils: The Place of Organic Matter in Sustaining Soils and Their Productivity, by Benjamin Wolf and George H. Snyder. Binghampton, NY: Food Products Press, an imprint of The Haworth Press, 2003.

"The Use of R.C.W. in Agriculture," by B. Noel. Translated from French at http://users.skynet.be/BRFinfo/anglais/abstract.htm.

Weeds

Weeds and What They Tell, by Ehrenfried E. Pfeiffer. Austin, TX: Bio-Dynamic Farming and Gardening, 1981.

Weeds and Why They Grow, by Jay L. McCaman. Austin, TX: Acres U.S.A., 1994.

Weeds: Control beyond Herbicides, by Harold Willis. Austin, TX: Acres U.S.A., 1993.

Weeds: Control Without Poisons, by Charles Walters. Austin, TX: Acres U.S.A., 1991; 2nd. ed., 1999.

Web Sites

Environmental Issues

Alternative Health News: www.altmedicine.org.

American PIE (American Public Information on the Environment). Publishes a free electronic newsletter called "Eco-Alert" on various environmental subjects. Contact Web site to subscribe. http: www.americanpie.org.

Ask Dr. Weil: www.pathfinder.com/drweil.

Audubon Action Alert. Free e-alert on environmental issues from the Audubon Society on decisions or regulations by government or corporations that hurt our families. Subscribe at http: audubon@audubon.org.

Beyond Pesticides. Database of information and a free electronic newsletter. www.beyondpesticides.org.

Chemical Injury Information Network (CIIN). A support and advocacy organization dealing with Multiple Chemical Sensitivities (MCS). It is run by the chemically injured for the benefit of the chemically injured, and focuses primarily on education, credible research into MCS, and the empowerment of the chemically injured. www.ciin.org.

Chemicals That Hurt Us. General information: www.chemicalbodyburden.com.

The Collaborative on Health and the Environment: www.cheforhealth.org.

DEAD ZONE—Science Museum of Minnesota. http://www.smm.org/deadzone/top.html.

Defenders of Wildlife. Free e-alert and newsletter on environmental issues affecting wildlife, national forests, public properties, and more. www.defenders.org.

Dirt Doctor: www.dirtdoctor.com.

"E" Magazine—The Environmental Magazine for consumers. www.emagazine.com.

Environmental Defense Fund (EDF). Environmental Defense is dedicated to protecting the environmental rights of all people, including future generations. Among these rights are clean air, clean water, healthy food, and flourishing ecosystems. They are guided by scientific evaluation of environmental problems, and the solutions they advocate will be based on science, even when it leads in unfamiliar directions. Free electronic newsletter; subscribe on website: www.edf.org.

Environmental News Network. A free daily electronic newspaper reporting on environmental news from around the world; presented in a headline-plus-summary format, click on headline for complete text. Customizable to certain issues. http: www.enn.com.

Environmental Working Group. Lots of information on various subjects. www.ewg.org.

ETC Group, or Action Group on Erosion, Technology and Concentration. ETC Group is dedicated to the conservation and sustainable advancement of cultural and ecological diversity and human rights. To this end, ETC Group supports socially responsible developments of technologies useful to the poor and marginalized and addresses international governance issues and corporate power (formerly the Rural Advancement Foundation International; RAFI). www.etcgroup.org.

Green Living: www.HowardGarrett.com.

Health Care Without Harm: www.noharm.org.

National Wildlife Federation. Online magazine of wildlife issues, political action alerts, and other issues affecting wildlife. Information on wildlife, from backyard habitat gardening to successful restoration projects, with lots of photos. www.nwf.org.

Northwest Coalition for Alternatives to Pesticides (NCAP). A free electronic newsletter on various issues; information sheets on the dangers of pesticides. www.pesticide.org.

PAN Pesticide Database. A one-stop location for current toxicity and regulatory information for pesticides, insecticides, herbicides, etc. This resource is a project of Pesticide Action Network of North America. www.pesticideinfo.org.

Pesticide Action Network: Advancing Alternatives to Pesticides. http: www.panna.org.

Pollution in Your Community. Get an in-depth pollution report for your county, covering air, water, chemicals, and more. Listing of toxic and/or polluted sites from EPA database, search by zip code or city. www.scorecard.org.

Rachel's Environment and Health Weekly. A free electronic newsletter on various subjects; well written, with extensive documentation and bibliography. Subscribe at http: www.rachel.org.

Texans for Alternatives to Pesticides (TAP). By educating the public and public officials about the harmful effects of pesticides and about safer alternatives, TAP seeks to promote its mission to "reduce the use of toxic pesticides in schools, homes, and public places." www.nopesticides.org/tapweb/indexshtm.

Texas Organic Research Center: www.texasorganicresearch.com.

Texas Pesticide Information Network (TXPIN)—Texas Consumers Union. Also a free electronic report on pesticide issues. Contact by e-mail at wongba@consumers.org or (512) 477-4431 or (512) 444-0811. www.texascenter.org/txpin.

TX PEER. Free electronic newsletter for Texas Public Employees for Environmental Responsibility. Texas PEER is a local chapter of a national alliance of local, state, and federal resource professionals. For Texas, subscribe at: www.txpeer.org.

Washington Toxics Coalition. Lots of useful information on toxic chemicals. www.watoxics.org.

Food-Related Sites

Eat Well Guide. This site allows one to enter one's zip code for a list of all restaurants and food sources within whatever mileage range you specify that sell or provide healthy food. www.eatwellguide.org.

Factory Farming: www.factoryfarming.com.

Farmers Markets. Listing of local farmers markets for fresh food. www.ams.usda.gov/farmersmarkets.

Organic Food Sources. Listing of local and organic food sources. www.localharvest.org.

Organic View. Excellent free electronic newsletter from the Organic Consumers Association (OCA) on food and health and safety issues—Campaign for Food Safety. Subscribe at http: www.organicconsumers.org.

Pollutants in Your Food. Free monthly e-mail bulletin to learn more about pesticides in produce. www.foodnews.org.

Real Milk. This site discusses the health benefits of real milk and lists sources. www.realmilk.org.

Safe to Use. Free electronic newsletter on various issues related to organic foods and environmental issues. www.safe2use.com.

Store Wars: www.storewars.org.

Sustainable Table: www.themeatrix.com.

Organic Agriculture and Gardening Sites

ATTRA. The National Sustainable Agriculture Information Service is managed by the National Center for Appropriate Technology (NCAT) and is funded under a grant from the USDA. Free weekly e-newsletter on the latest research projects, success stories, etc., as related to sustainable agriculture (i.e., organic) and related environmental issues from business to politics and government regulation. www.attra.ncat.org.

Biological Control News. Free electronic news magazine. www.entomology.wisc.edu/mbcn.mbcn.html.

Gardening information by Malcolm Beck, founder of Gardenville Organic Fertilizer and Compost Company: www.malcolmbeck.com.

Gardening information for the Houston Gulf Coast: www.urban-nature.org.

Growing Solutions: www.growingsolutions.com.

Henry Wallace Institute for Alternative Agriculture. Publishes the *Journal of Sustainable Agriculture*. www.hawiaa.org.

Howard Garrett's Basic Organic Gardening Program. Information on how to grow plants, organically prevent pests, and solve gardening problems in a safe and natural manner. www.dirtdoctor.com.

International Institute for Sustainable Agriculture: www.permaculture-institute.com.

"Lazy Gardener by Brenda Smith." Information on Houston gardening events (go to features section and click on Brenda's Garden). www.guidrynews.com or http://blogs.chron.com/lazygardener/.

Natural Solutions: www.nsei.net.

The New Farm. Rodale Research Institute free electronic organic farming newsletter. www.newfarm.com.

Organic Farming Research Foundation. Publishes a free electronic newsletter called SCOAR (Scientific Congress on Organic Agricultural Research). www.ofrf.org.

Organic gardening information for the Houston Gulf Coast: www.watersmart.cc.

Organic Horticulture Business Alliance (OHBA). OHBA is set up as a committee of Urban Harvest and is designed for the advancement of the landscape professionals in the metro Houston area. www.ohbaonline.org.

Organic Materials Review Institute. The institute provides certifiers, growers, manufacturers, and suppliers an independent review of products intended for use in certified organic production, handling, and processing. www.omri.org.

Rodale Press. Publisher of many good books on organic and natural techniques. www.rodalepress.com.

Southern Sustainable Agriculture Working Group (SSAWG). Information and research on sustainable agriculture for the southern states. www.ssawg.org.

Texas Organic Farmers and Gardeners Association. Our new site will be designed to help promote local farms, ranches, food co-ops, local-food restaurants, farmers markets, and organic suppliers throughout Texas. www.tofga.org.

Urban Harvest. Houston's Community Gardening program, offers classes on organic gardening and related subjects. www.urbanharvest.org.

Societies

American Bamboo Society, 750 Krumkill Road, Albany, NY 12203. www.bamboo.org.

American Hibiscus Society. Lone Star Chapter: www.lonestarahs.org; Space City Chapter: www.spacecityahs.org.

American Rose Society: www.ars.org.

Bromeliad Society: www.bromeliadsocietyhouston.org.

Gulf Coast Fruit Study Group: www.harris-tx.tamu.edu/hort/fruit.

Herb Society of America: www.herbsociety-stu.org.

American Bonsai Society: www.abonsai.org
American Camellia Society: www.camellias-acs.com
American Fern Society: www.amerfern.org
American Rose Society: www.ars.org
Cactus and Succulent Society of America: www.cssainc.org
International Palm Society: www.palms.org
International Bulb Society (formerly American Amaryllis Society: www.bulbsociety.org
International Oleander Society: www.oleander.org.
International Plumeria Society (Houston): www.plumeria.org.
Lone Star Daylily Society: www.lonestardaylilysociety.org.
National Orchid Society: www.ncos.us/ncos/
Native Plant Society of Texas, Houston: www.npsot.org/houston.
Texas Rose Rustlers: www.texasroserustlers.org.
Urban Harvest (Community Gardening Program). Excellent source for organic gardening information and classes/seminars. Also offers advanced training for Master Gardeners, (713) 880-5540: www.urbanharvest.org.

Index

Page numbers in italics indicate images.